AF389154

Low-power Wearable Healthcare Sensors

Low-power Wearable Healthcare Sensors

Special Issue Editors

R. Simon Sherratt
Nilanjan Dey

MDPI • Basel • Beijing • Wuhan • Barcelona • Belgrade

Special Issue Editors

R. Simon Sherratt
University of Reading
UK

Nilanjan Dey
Techno India College of Technology
India

Editorial Office
MDPI
St. Alban-Anlage 66
4052 Basel, Switzerland

This is a reprint of articles from the Special Issue published online in the open access journal *Electronics* (ISSN 2079-9292) from 2018 to 2020 (available at: https://www.mdpi.com/journal/electronics/special_issues/lowpower_sensors).

For citation purposes, cite each article independently as indicated on the article page online and as indicated below:

LastName, A.A.; LastName, B.B.; LastName, C.C. Article Title. *Journal Name* **Year**, *Article Number*, Page Range.

ISBN 978-3-03936-479-4 (Hbk)
ISBN 978-3-03936-480-0 (PDF)

Contents

About the Special Issue Editors

R. Simon Sherratt (Professor) received a B.Eng. degree in electronic systems and control engineering from Sheffield City Polytechnic, Sheffield, U.K., in 1992, a M.S. degree in data telecommunications, and a Ph.D. degree in video signal processing from the University of Salford, Salford, U.K., in 1994 and 1996, respectively. In 1996, he was appointed as a lecturer in electronic engineering in the University of Reading, Reading, U.K., where he is currently Professor of biosensors and leads the wearable devices research activities. His research interests include signal processing and personal communications in consumer devices focusing on wearable devices and healthcare, primarily for the residential management of Parkinson's disease and the automatic detection of human emotion. He is a Fellow of the IET and a Fellow of the IEEE.

Nilanjan Dey is Assistant Professor in the Department of Information Technology at Techno International New Town (formerly known as Techno India College of Technology), Kolkata, India. He has also served as honorary Visiting Scientist at Global Biomedical Technologies Inc., CA, USA (2012–2015). He was awarded his Ph.D. from Jadavpur University in 2015. He is the Editor-in-Chief of the International Journal of Ambient Computing and Intelligence, IGI Global. He is the Series Co-Editor of Springer Tracts in Nature-Inspired Computing, Springer Nature; Series Co-Editor of Advances in Ubiquitous Sensing Applications for Healthcare, Elsevier; and Series Editor of Computational Intelligence in Engineering Problem Solving and Intelligent Signal processing and data analysis, CRC. He has authored/edited more than 50 books with Springer, Elsevier, Wiley, and CRC Press and published more than 300 peer-reviewed research papers. His main research interests include medical imaging, machine learning, computer-aided diagnosis, data mining, etc. He is the Indian Ambassador of the International Federation for Information Processing (IFIP)—Young ICT Group.

Editorial

Low-Power Wearable Healthcare Sensors

Robert Simon Sherratt [1,*] and Nilanjan Dey [2]

1   Department of Biomedical Engineering, University of Reading, Reading RG6 6AY, UK
2   Techno India College of Technology, Kolkata, West Bengal 700156, India; neelanjan.dey@gmail.com
*   Correspondence: r.s.sherratt@reading.ac.uk

Received: 21 May 2020; Accepted: 25 May 2020; Published: 27 May 2020

1. Introduction

Medical science has taken great steps to enable us to live longer and healthier lives. While hospitals are vital for intervention-based healthcare, hospital care is expensive, increasing in cost, and with the continual increase in the global elderly population, we need solutions to enable people to stay healthy. While smart home technology solutions are enabling people to live in their homes for longer, research has shown that more personalized data are needed to improve services and decisions, and this is where wearable devices are proving their worth. Sensors can be placed on and around the body, in clothing, in shoes, in jewelry, and in many other accessories to measure movement, physiology, environment, and even mood/emotion. Such technology will become more common, and indeed vital, in long-term health monitoring. Perhaps the real potential of such devices is not just to monitor, but to have interactive communication with cloud services to offer personalized and ongoing real-time healthcare advice, enabling people to manage their health and to reduce hospital admissions. Therefore, the challenge of the next generation of wearable healthcare devices is to offer a wide range of sensing, computing, communication, and human–computer interactions, all within a tiny device with limited resources and electrical power.

2. Low-Power Wearable Healthcare Sensors

The aim of this Special Issue was to highlight the research challenges being tackled to enable the next generation of low-power wearable healthcare sensors. Six interesting and thought-provoking papers have been selected which present a wide range of challenges and solutions:

1.   Enhancements to software to manage and constrain peak electrical current consumption in a wearable healthcare device [1],
2.   A wearable low-power insulin delivery method for people with diabetes [2],
3.   The design of an Internet of Things (IoT)-based sensor node for eHealth [3],
4.   Human joint angle estimation and movement analysis using polymer optical fiber curvature sensors and inertial measurement units [4],
5.   A neural spike detector system as an interesting roadmap to wearable devices monitoring brain activity [5],
6.   CMOS Interfaces for the detection of biological fluids (sweat, tears, saliva, and urine) as part of a connected internet-of-wearables [6].

We hope you enjoy reading this Special Issue and are inspired to address the technological challenges to help prolong human lives.

3. Conclusions

Wearable devices have become a must-have consumer device. Such devices are commonly seen in healthcare, fitness, and location-tracking applications, but this is just the start. The technology behind wearable devices is now maturing to the point where consumer-grade devices can be used to help people manage chronic conditions, including Parkinson's disease, stroke, diabetes, and dementia. However, for reliable deployment at home, wearable devices must now balance the need to operate for a long time between charges with having ever increasing computational and communicative resources.

Research to enable wearable devices to last longer per charge is extremely important because, as these devices are almost exclusively powered by batteries, their size is limited for user comfort and usability. Devices that last a long time per charge enable the long-term monitoring of the user for improved health decisions. While battery capacity can be improved, research also indicates that smarter and lower-power CPU architectures, improved wireless connectivity, and task scheduling can also enable devices to last longer. In the future, wearable devices will not only monitor movement and heartrate, but also incorporate sensors to capture the users' biological states, as we are already now seeing for emotion and stress [7].

Author Contributions: Conceptualization, R.S.S. and N.D.; methodology, R.S.S. and N.D.; writing—review and editing and project administration, R.S.S. and N.D. All authors have read and agreed to the published version of the manuscript.

Funding: This research received no external funding.

Conflicts of Interest: The authors declare no conflict of interest.

References

1. Sherratt, R.S.; Janko, B.; Hui, T.; Harwin, W.S.; Dey, N.; Díaz-Sánchez, D.; Wang, J.; Shi, F. Task Scheduling to Constrain Peak Current Consumption in Wearable Healthcare Sensors. *Electronics* **2019**, *8*, 789. [CrossRef]
2. Berián, J.; Bravo, I.; Gardel, A.; Lázaro, J.L.; Hernández, S. A Wearable Closed-Loop Insulin Delivery System Based on Low-Power SoCs. *Electronics* **2019**, *8*, 612. [CrossRef]
3. Petrellis, N.; Birbas, M.; Gioulekas, F. On the Design of Low-Cost IoT Sensor Node for e-Health Environments. *Electronics* **2019**, *8*, 178. [CrossRef]
4. Valencia-Jimenez, N.; Leal-Junior, A.; Avellar, L.; Vargas-Valencia, L.; Caicedo-Rodríguez, P.; Ramírez-Duque, A.A.; Lyra, M.; Marques, C.; Bastos, T.; Frizera, A. A Comparative Study of Markerless Systems Based on Color-Depth Cameras, Polymer Optical Fiber Curvature Sensors, and Inertial Measurement Units: Towards Increasing the Accuracy in Joint Angle Estimation. *Electronics* **2019**, *8*, 173. [CrossRef]
5. Vallicelli, E.A.; Reato, M.; Maschietto, M.; Vassanelli, S.; Guarrera, D.; Rocchi, F.; Collazuol, G.; Zeitler, R.; Baschirotto, A.; De Matteis, M. Neural Spike Digital Detector on FPGA. *Electronics* **2018**, *7*, 392. [CrossRef]
6. Dei, M.; Aymerich, J.; Piotto, M.; Bruschi, P.; del Campo, F.J.; Serra-Graells, F. CMOS Interfaces for Internet-of-Wearables Electrochemical Sensors: Trends and Challenges. *Electronics* **2019**, *8*, 150. [CrossRef]
7. Zamkah, A.; Hui, T.; Andrews, S.; Dey, N.; Shi, F.; Sherratt, R.S. Identification of suitable biomarkers for stress and emotion detection for future personal affective wearable sensors. *Biosensors* **2020**, *10*, 40. [CrossRef] [PubMed]

 *electronics*

Article

Task Scheduling to Constrain Peak Current Consumption in Wearable Healthcare Sensors

Robert Simon Sherratt [1,*], Balazs Janko [1], Terence Hui [2], William S. Harwin [1], Nilanjan Dey [3], Daniel Díaz-Sánchez [4], Jin Wang [5] and Fuqian Shi [6]

[1] Department of Biomedical Engineering, University of Reading, Reading RG6 6AY, UK
[2] IO Senses Ltd, Reading RG1 3BJ, UK
[3] Department of Information Technology, Techno India College of Technology, West Bengal 740000, India
[4] Department of Telematic Engineering, Universidad Carlos III de Madrid, Leganés, 28911 Madrid, Spain
[5] School of Computer & Communication Engineering, Changsha University of Science & Technology, Changsha 410114, China
[6] College of Information and Engineering, Wenzhou Medical University, Wenzhou 325035, China
* Correspondence: r.s.sherratt@reading.ac.uk; Tel.: +44-118-378-8588

Received: 30 May 2019; Accepted: 12 July 2019; Published: 15 July 2019

Abstract: Small embedded systems, in our case wearable healthcare devices, have significant engineering challenges to reduce their power consumption for longer battery life, while at the same time supporting ever-increasing processing requirements for more intelligent applications. Research has primarily focused on achieving lower power operation through hardware designs and intelligent methods of scheduling software tasks, all with the objective of minimizing the overall consumed electrical power. However, such an approach inevitably creates points in time where software tasks and peripherals coincide to draw large peaks of electrical current, creating short-term electrical stress for the battery and power regulators, and adding to electromagnetic interference emissions. This position paper proposes that the power profile of an embedded device using a real-time operating system (RTOS) will significantly benefit if the task scheduler is modified to be informed of the electrical current profile required for each task. This enables the task scheduler to schedule tasks that require large amounts of current to be spread over time, thus constraining the peak current that the system will draw. We propose a solution to inform the task scheduler of a tasks' power profile, and we discuss our application scenario, which clearly benefited from the proposal.

Keywords: wearable; low-power; embedded; task scheduler; healthcare

---

1. Introduction

The use of personalized healthcare devices, from wearable fitness trackers to smartphone apps, has been of growing significance. We are now seeing artificial intelligence (AI)-based intelligent devices address the daily challenges of those living with chronic diseases, including Alzheimer's, cerebral palsy, chronic obstructive pulmonary disease (COPD), diabetes, hypertension, multiple sclerosis, obesity and Parkinson's disease. One of the main challenges has been to embed enough computing power into a wearable device while keeping the device lightweight. This places demanding constraints on the battery weight and size, so power management is of utmost importance.

Power efficiency has been a long-running research challenge, and is often addressed though low-power designs enabling a device to run for a long time between charges. This is particularly important for devices aimed at addressing the needs of people living with dementia who may not remember to charge the device.

To enable battery-powered devices to run for a long time between charges, research has focused on hardware solutions, software solutions, and hybrids of the two. Achieving a low-power solution is often seen as a balance between reducing battery current by placing the central processing unit (CPU) into a low-power idle state (i.e., sleep mode) versus the need for a CPU to consume battery current to execute task code at full clock speed. While there are numerous smart devices on the market based on common operating systems, these generic and ubiquitous devices are designed to support many features, and as such, they have a relatively heavy use of power. On the other hand, research has indicated that low-power embedded solutions can come from devices that are based on a real-time operating system (RTOS) that has been targeted to support the resource-constrained embedded device with their application solely focused on the use case.

An important element of an RTOS is the task scheduler (TS), which is responsible for deciding when software tasks are to be executed (preemptively), and is typically based on an assigned task priority [1]. In addition, power management is also an important design enabler for an RTOS. Modern RTOS systems address this objective by only waking up the CPU when processing is required, and enter a low-power idle state when all tasks have completed or are waiting. While dynamic voltage frequency scaling (DVFS) and dynamic power management (DPM) are important and topical research issues [2–4], our focus in this paper is concerned with small embedded devices that typically can only be either running or idle, and only use one core voltage. It has been also shown that when the TS places a system into idle, the battery current is reduced, enabling certain battery chemistries to recover some charge [5], potentially up to 20%. Therefore, such duty cycling is easily enabled using RTOS mechanisms.

Li et al. [6] presented a dynamic task scheduling algorithm focusing on energy reduction. The work considered the dynamic scheduling of algorithms to obtain a global minimization of power in a cloud computing platform.

Liu et al. [7] considered a delay optimal task scheduling technique on a mobile device to minimize the power of a local CPU load or to offload a task to a cloud edge server. Computation tasks were scheduled based on the queueing state of a task buffer, the execution state of the CPU, and the state of the transmission unit. Their algorithm searched for an optimal solution that minimizes the global power consumption, but of course, the search itself consumed power.

Ghofrane et al. [8] used a neural network (NN) to obtain a global minimized power schedule, while Li and Wu [9] presented an ant colony optimized (ACO) search method for their task scheduler. While both approaches produce global power savings, the application of such schemes are directed toward higher-end computing systems rather than wearable healthcare devices due to the considerable power needed to perform the optimizing search.

Nguyen et al. [10] presented interesting work by considering various algorithms for task scheduling for the Internet of Things (IoT) by considering completion time and operating costs. While the work itself was not concerned with power scheduling, it is possible that the concepts presented may also be used for power management.

Ahmad et al. [11] presented a formal method for the evaluation of real-time task scheduling in mission critical systems. Such research will be useful to medical IoT systems, but they did not consider power management.

Zagan et al. [12] presented scheduling of tasks via the use of a hardware architecture as opposed to a software-based RTOS.

Singh et al. [13] considered task scheduling for data centers by considering process time flow in order to minimize the overall energy consumed by the data center. The work presented an interesting power model, but the focus of the work was to minimize overall energy, not constrain peak current.

Ahmad et al. [14] noted that as small embedded devices have limited resources (power and memory), then an RTOS should consider the required tasks to produce minimum CPU power that still gives correct operation. The work presented a resource-aware approach to minimize tasks based on device profiles. A power profile for each device is an interesting idea, but again, the focus of the work was the global minimization of CPU power, not to constrain peak current.

Chowdhury and Chakrabarti [15] presented interesting work by considering power profiles for battery-operated systems in order to maximize the residual charge left in the battery. Their system used dynamic voltage scaling (DVS), which is currently only present on high-end CPUs, and not applicable in our case with the current technology level. Again, this work was useful to create overall power savings, but did not constrain peak current demands.

It is clear from the literature that where electrical power is concerned, the previous research objectives have been to schedule tasks to globally minimize the consumed power, not to constrain peak power. To the authors' knowledge, no work has considered task scheduling for the context of constraining peak electrical current. This position paper proposes that low-power embedded systems, particularly wearable healthcare devices where battery operation is paramount, can significantly benefit when the software task scheduler embedded in the operating system is informed of the current drain for each task, so that higher-current tasks can be separated and executed over time in order to guarantee a maximum peak current drain. Constraining the peak current aids in the design of the internal regulator and battery management circuits. It also enables energy harvesting systems and brownout recovery due to lower peaks of current in the device's startup condition. In this paper, we present how the need for such a system was realized through our real-life application scenario, and then present our general proposal to address the issue.

The paper is organized as follows: Section 2 discusses the development of the concept of the need for power-aware schedulers to reduce peak power. Section 3 presents our proposed solution, Section 4 presents a discussion, and Section 5 presents our conclusions.

2. Application Scenario

The position statement from this paper is derived from experience in creating the wearable device used in the SPHERE (https://www.irc-sphere.ac.uk) project. SPHERE was a large interdisciplinary research project to create technology and algorithms for monitoring people for residential healthcare with the objective of enabling people to live in their home for longer. The SPHERE architecture integrates environmental, video, and wearable sensors into a smart hub and backbone network [16] enabling intelligent processing and data-driven decisions from the fusion of the sensor data and data mining.

The SPHERE wearable [16] was specifically designed for low-power operation to enable pervasive monitoring; otherwise, behaviors could be influenced. It was based on a Texas Instruments CC2650 (Dallas, TX, USA) [17] system on chip (SoC) supporting either Bluetooth 4.2 [18] (often called Bluetooth low energy (BLE) or Bluetooth smart) or 802.15.4 [19]. The SoC contains an ARM Cortex-M0 CPU dedicated to controlling just the radio interface and an ARM Cortex-M3 CPU for hosting the application code. The SPHERE wearable also contained a FLASH memory device [20] for local permanent storage, two low-power accelerometers [21] allowing for gyro-free 3D spatial measurements [22], and an inertial measurement unit (IMU) [23]. The CC2650 does have an internal DC–DC convertor in the SoC, allowing the SoC to be powered from 1.8 V–3.8 V. However, a high performance and efficient external buck convertor [24] was used to step down the voltage from a 100-mA lithium polymer (LiPo) battery (typically 4.2 V to 3.6 V) to 1.8 V for the SoC core. The ability to wirelessly charge the LiPo battery via Qi was also included. The wearable was programed using an RTOS [25].

The SPHERE wearable was configured to use the BLE radio in a non-connected mode [26]. The wearable application packaged the sensed movement, battery voltage, and SoC core temperature measurements into a non-connectable BLE advertisement packet (ADV), which was detected by passive BLE hubs spread across the residence. In BLE, devices advertise their presence by transmitting on the advertising channels (37, 38, and 39) in sequence [26]. Room hubs capture the encapsulated data and forward the data to be fused with the video data. Typically, the wearable lasted for four weeks on one charge when transmitting one BLE ADV packet per second, while variants of more frequent transmissions were also obtained. The SPHERE wearable in the above configuration worked well when deployed.

Then, the SPHERE wearable was required to be used in a different set of experiments on monitoring people with Parkinson's disease in their own home continually for three days, with the aim of capturing movement events called near-falls [27], which are signature movements that occur prior to an actual fall. This application required the IMU to continually sample movement at 60 Hz and store the data to the local FLASH for download three days later. An interrupt was generated by the IMU 60 times a second to request that the CPU read the IMU's data. As the battery current when writing a single page of data to the FLASH was 12.4 mA (SoC core current plus FLASH page write current), then the movement data was not in this case streamed over BLE in order to save power, enabling three-day operation.

The battery current profile when reading data from the IMU and writing the data to a FLASH page is given in Figure 1 over the whole IMU interrupt service routine (ISR). In Figure 1, point (a) indicates that before the IMU interrupt, the CPU is in its idle state; therefore, the 1.2-mA wearable device battery current is dominated by the IMU, which is always on. When the interrupt from the IMU occurs, the SoC changes from its low-frequency (LF) clock to the high-frequency (HF) clock at point (b). Once the HF clock is running, the CPU is fully operational to execute code using the HF clock with the battery current increasing to 5.2 mA. The data in the IMU is read via the SoC SPI bus at point (c), post-processed at point (d), and then, the data is written to local FLASH memory at point (e), resulting in a significant peak of battery current (12.4 mA). Then, the SoC returns back to idle while the IMU is still running at point (f).

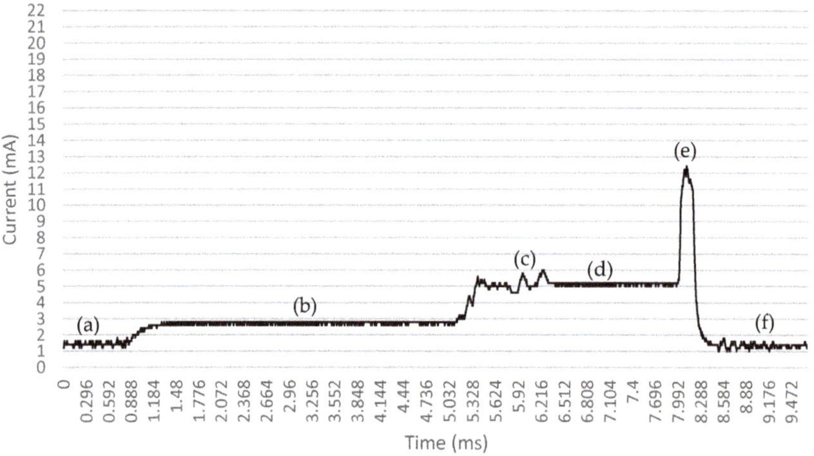

Figure 1. Current profile of the wearable over the inertial measurement unit (IMU) interrupt service routine (ISR). (a) is the idle current with IMU always on (1.2 mA); point (b) is where the system on chip (SoC) wakes up to process the ISR (2.8 mA). Then, the SoC enables the high-frequency clock; point (c) is where the SoC reads the IMU data over the SPI bus, and point (d) is where the IMU data is post-processed (5.2 mA); point (e) is where the IMU data is stored in the FLASH memory using a page write process (12.4 mA); and point (f) is where SoC then returns to idle with just the IMU running.

In our near-fall detection application, a connectable BLE mode was used to enable a smartphone/tablet to control the wearable over BLE via an app enabling the user to start/stop the logging, and to download the movement data. Therefore, the wearable advertised its presence using a connectable BLE ADV process once per second, requiring the transmissions of three peaks from 10.5 mA to 12 mA, as shown in Figure 2, over point (g). A user can connect to the device to start the logging process, and then disconnect, leaving the device logging to the FLASH memory over three days while the device periodically advertises its presence for a host to connect to the device.

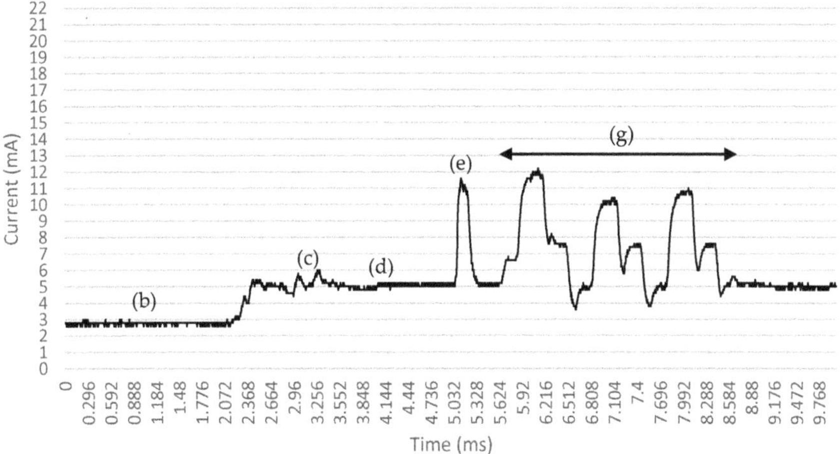

Figure 2. Current profile of just the latter part of the IMU ISR followed by a Bluetooth low energy (BLE) advertisement packet (ADV) process. Point (g) is the three transmissions on each of the three BLE ADV frequencies. Of note is that the FLASH write process point (e) and the BLE ADV process point (g) are separated in time, extending the current drain into that of period (f), as depicted in Figure 1.

However, it was found that the device would brownout after 24 h of successful operation. After investigations, we realized that the ADV transmission process was asynchronous to the operation of the FLASH page write process. When both the BLE ADV and the FLASH write processes occurred at the same time, then the current drawn from the Buck convertor contained the SoC core current, the current to write to the FLASH, and the current required for the BLE ADV transmission, as can be seen in Figure 3, point (h). By using the external buck to drive the SoC core at its required 1.8 V in order to minimize power losses, then the consequence was that when a large instantaneous current is drawn by the wearable, then the Buck convertor took time to respond, and the SoC core voltage dropped, causing the brownout.

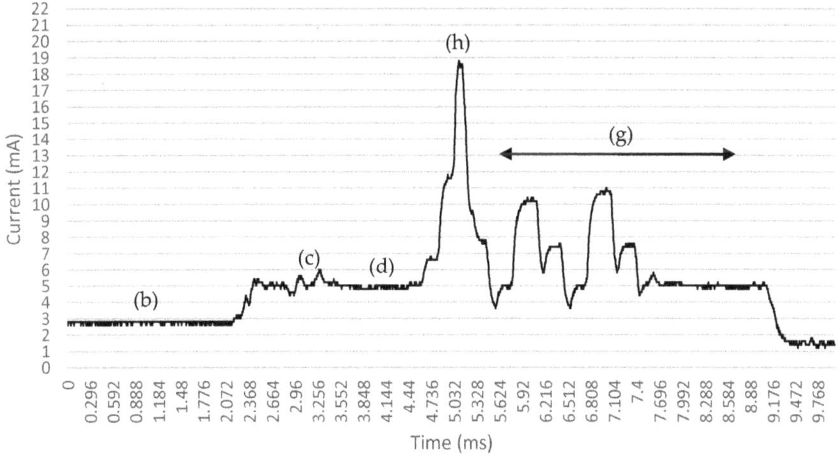

Figure 3. Current profile of just the latter part of the IMU ISR where the BLE ADV process occurs over the same time as the first BLE ADV transmission. Point (h) is now the summation of the SoC current, the FLASH write current, and the BLE ADV transmit current (18.8 mA).

Application Scenario Solution

The issue of the brownout, in this case, was simply solved by synchronizing the ADV transmission process with the FLASH write process, such that the ADV transmission occurred spaced in-between successive FLASH write processes, allowing the buck time to recover. Thus, it removed the large peak of current by spreading the current drawn from the buck over time.

It was these experiments and solutions that have led to our position in this paper of proposing task schedulers in embedded systems to be aware of the current (power) requirements for each task, and for the scheduler to schedule tasks based not only on task priority, but also to separate in time tasks that have large peak currents, thus constraining the maximum current drawn at any time. In effect, this process associated a peripheral's electrical current profile with a given software task. Such an approach does not itself save any power; rather, it constrains the peak current drawn.

3. Proposed Modifications to RTOS Task Parameters

From the literature, it was clear that while there are many works that have focused on the creation of task schedulers to schedule tasks with the objective of globally minimizing electrical power, there are no works that have focused the task scheduler to constrain peak electrical current. The placement of such a constraint would not only simplify the power delivery circuits of wearable devices, but also have advantages in EMI reduction.

While Ahmad et al. [14] proposed that an RTOS should consider the required tasks to produce minimum global CPU power via a resource-aware approach based on device profiles, we propose that a scheduler can be informed of the required current consumption for each task when the task is instantiated to enable the task scheduler to then separate higher current tasks over time.

The design and modification of operating system task schedulers is outside of the scope of this work. Rather, we propose the position that there are significant advantages to be gained in the power delivery of wearable devices if task schedulers were made aware of the current required for each task. In order to create such a task scheduler, then the task scheduler would need to be informed of an expected electrical current profile for each software task. Clearly, a software task itself requires current for the CPU to perform the task, but may also use a peripheral device (e.g., the case of a FLASH memory page write) and the associated current to perform that task. In the case where the required peripheral current may be of relatively short duration to the overall software task, then the task can be split into multiple concurrent tasks, as was the application scenario presented in Section 2.

In order to create software tasks, each task needs to be instantiated in runtime though the scheduler. The typical task instantiation code is given in Figure 4. It can be seen in Figure 4 that a structure taskParams is created to contain the task parameters required, including the task stack TaskStack (of size TASK_STACK_SIZE) and task priority. Then, the task scheduler can construct the new task in the scheduler in runtime using the information in taskParams and the address of the task.

```
Task_Struct Task;
Char TaskStack[TASK_STACK_SIZE];
Task_Params taskParams;

// Configure task
Task_Params_init(&taskParams);
taskParams.stack = TaskStack;
taskParams.stackSize = TASK_STACK_SIZE;
taskParams.priority = 1;

Task_construct(&Task, taskFxn, &taskParams, NULL);
```

Figure 4. Typical method in a real-time operating system (RTOS) to instantiate task taskFxn with the task parameters configured in a structure.

As parameters are used to define the task, then an additional parameter, which was in this paper called the additional_current, may be added to the structure to pass the expected additional current (of defined units) that needs to be taken into consideration when the task runs (due to the peripheral or radio transceiver action), as shown in Figure 5. In this way, tasks can then be scheduled with the full knowledge of the required current that will be drawn when that task is ran, thus allowing for the scheduling of tasks to constrain the peak current. The FLASH write task sends a command to the FLASH device to start its internal process of transferring the page memory into the FLASH memory, and it is when this transfer process occurs that the FLASH draws a large current. Hence, the FLASH write task has been deliberately kept to a short duration so that we can approximate the FLASH write task to have constant current (the FLASH write current) across the whole of the task. A compile-time directive can be used to indicate the acceptable peak current, or target constrained peak current for a given system, with the scheduler arranging the tasks to meet the target constrained peak current. In the case of the scheduler being unable to schedule the tasks given the peak current constraint, then a runtime error can be generated in the same way as a RTOS running out of memory space to manage new instantiated tasks.

```
Task_Struct Task;
Char TaskStack[TASK_STACK_SIZE];
Task_Params taskParams;

// Configure task
Task_Params_init(&taskParams);
taskParams.stack = TaskStack;
taskParams.stackSize = TASK_STACK_SIZE;
taskParams.priority = 1;
taskParams.additional_current = 12;

Task_construct(&Task, taskFxn, &taskParams, NULL);
```

Figure 5. Modified method to instantiate task taskFxn with the addition of new parameter additional_current to inform the task scheduler of the required additional current (in this case 12 mA) that will be drawn when that task runs.

In the specific case of Figure 3, where the FLASH write current occurred at the same time as the BLE ADV transmission, we applied the concept of the scheduler to delay the FLASH write process, and the results are presented in Figure 6. As can be seen in Figure 6, the FLASH write process, point (e), has been delayed in time to separate its current from the BLE ADV transmission, point (g), thus successfully constraining the peak current drawn from the battery.

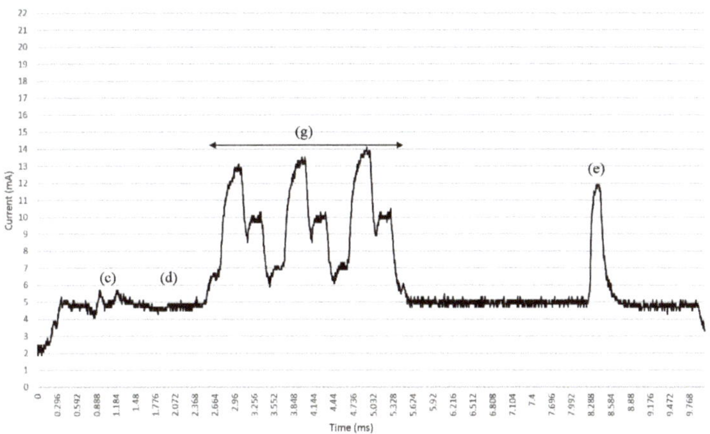

Figure 6. Use of the scheduler concept to delay the FLASH write current to run after the BLE ADV transmission has occurred. Point (e) has now been delayed 5 ms to after the start of point (g), separating the large current into two processes constraining the peak current.

4. Discussion

Through the application scenario that we have experienced, and the results presented in this paper, we have formed a position statement to propose that there are significant advantages to be gained when modifying embedded RTOS task schedulers to further consider the electrical current required to process each task (including peripherals and radio transceivers) when performing the task scheduling process. It is not just a case of scheduling to minimize power, but also to constrain peak power. Clearly, the past work has been very successful regarding deciding when to run tasks in order to minimize the overall electrical current. Such an approach is very useful to enable battery-powered devices to last longer. However, tiny healthcare sensors are constrained by not only their overall available current, but also by their ability to supply large peaks of current due to constrained batteries and voltage regulators. The ability to spread the current drawn over time, rather than to have a large current peak, has significant advantages to simplify power systems that drive wearable healthcare devices.

It is not surprising that in many application scenarios, there are peaks of activity over time (e.g., sampled sensor data) that result in large peaks of current. However, as presented in this paper, the scheduler can delay tasks to a point in the future to constrain peak current while not actually consuming any more current.

The implications of a power-aware scheduler with the ability to constrain electrical current are significant. Not only can such technology be used in smart watches and smart phones, but it can also be used in the control of tasks in smart cars, where considerably large peaks of current can be present.

5. Conclusions

This paper has considered the use of task scheduling in the context of constraining peak electrical current in embedded systems, particularly power sensitive wearable healthcare devices. We have shown that it is possible that such devices can, at a given time, draw peak currents leading to short-term peak demands on batteries and power circuits, ultimately causing supply voltage drops and brownouts, and difficulties in energy harvesting systems.

We propose that an RTOS task scheduler should consider not only the task priority to decide when to schedule tasks, but also the power (current) demands of each task. To avoid large peaks in current, this research work associated an electrical current profile to a software task. Therefore, the objective of the scheduler is to separate over time tasks requiring large current drains rather than schedules to reduce the global power drain. Our task scheduler can delay tasks that have large current profiles so as to constrain the current drawn from the supply within a given limit.

By constraining the peak power current that a device may draw, this then enables simpler power supply designs, energy harvesting, and can reduce electromagnetic interference effects from large pulses of current.

Author Contributions: Conceptualization, R.S.S., B.J., T.H., W.S.H. and D.D.-S.; methodology, N.D., F.S. and J.W.; software, R.S.S.; validation, B.J.; investigation, R.S.S.; writing—original draft preparation, R.S.S.; writing—review and editing, R.S.S., W.S.H., N.D. and T.H.

Funding: This research was partially funded by the UK Engineering and Physical Sciences Research Council (EPSRC), grant number EP/K031910/1, and in part by the Royal Society and National Natural Science Foundation of China International Exchanges 2017 Cost Share (China), grant number IEC\NSFC\170363. The APC was kindly funded by MDPI as part of the special issue on Low-power Wearable Healthcare Sensors.

Conflicts of Interest: The authors declare no conflict of interest.

References

1. Singh, S.; Chana, I. A Survey on scheduling in cloud computing: Issues and challenges. *J. Grid Comput.* **2018**, *14*, 217–264. [CrossRef]
2. Bambagini, M.; Marinoni, M.; Aydin, H.; Buttazzo, G. Energy-aware scheduling for real-time systems: A survey. *ACM Trans. Embed. Comput. Syst.* **2016**, *15*, 7. [CrossRef]

3. Lin, X.; Wang, Y.; Xie, Q.; Pedram, M. Task scheduling with dynamic voltage and frequency scaling for energy minimization in the mobile cloud computing environment. *IEEE Trans. Serv. Comput.* **2015**, *8*, 175–186. [CrossRef]

4. Li, X.; Xie, N.; Tian, X. Dynamic voltage-frequency and workload joint scaling power management for energy harvesting multi-core WSN node SoC. *Sensors* **2017**, *17*, 310. [CrossRef] [PubMed]

5. Arora, H.; Sherratt, R.S.; Janko, B.; Harwin, W. Experimental validation of the recovery effect in batteries for wearable sensors and healthcare devices discovering the existence of hidden time constants. *J. Eng.* **2017**, *2017*, 548–556. [CrossRef]

6. Li, Y.; Chen, M.; Dai, W.; Qiu, M. Energy optimization with dynamic task scheduling mobile cloud computing. *IEEE Syst. J.* **2015**, *11*, 96–105. [CrossRef]

7. Liu, J.; Mao, Y.; Zhang, J.; Letaief, K.B. Delay-optimal computation task scheduling for mobile-edge computing systems. In Proceedings of the 2016 IEEE International Symposium on Information Theory (ISIT), Barcelona, Spain, 10–15 July 2016. [CrossRef]

8. Ghofrane, R.; Gharsellaoui Hamza, G.; Samir, B.A. New optimal solutions for real-time scheduling of reconfigurable embedded systems based on neural networks with minimisation of power consumption. *Int. J. Intell. Eng. Inform.* **2018**, *6*. [CrossRef]

9. Li, G.; Wu, Z. Ant colony optimization task scheduling algorithm for SWIM based on load balancing. *Future Internet* **2019**, *11*, 90. [CrossRef]

10. Nguyen, B.H.; Binh, H.T.T.; Anh, T.T.; Son, D.B. Evolutionary algorithms to optimize task scheduling problem for the IoT based bag-of-tasks application in cloud–fog computing environment. *Appl. Sci.* **2019**, *9*, 1730. [CrossRef]

11. Ahmad, S.; Malik, S.; Ullah, I.; Park, D.-H.; Kim, K.; Kim, D. Towards the design of a formal verification and evaluation tool of real-time tasks scheduling of IoT applications. *Sustainability* **2019**, *11*, 204. [CrossRef]

12. Zagan, I.; Găitan, V.G. Hardware RTOS: Custom scheduler implementation based on multiple pipeline registers and MIPS32 architecture. *Electronics* **2019**, *8*, 211. [CrossRef]

13. Singh, P.; Khan, B.; Vidyarthi, A.; Alhelou, H.H.; Siano, P. Energy-aware online non-clairvoyant scheduling using speed scaling with arbitrary power function. *Appl. Sci.* **2019**, *9*, 1467. [CrossRef]

14. Ahmad, S.; Malik, S.; Ullah, I.; Fayaz, M.; Park, D.-H.; Kim, K.; Kim, D. An Adaptive approach based on resource-awareness towards power-efficient real-time periodic task modeling on embedded IoT devices. *Processes* **2018**, *6*, 90. [CrossRef]

15. Chowdhury, P.; Chakrabarti, C. Static task-scheduling algorithms for battery-powered DVS systems. *IEEE Trans. Very Large Scale Integr. Syst.* **2005**, *13*, 226–237. [CrossRef]

16. Fafoutis, X.; Vafeas, A.; Janko, B.; Sherratt, R.S.; Pope, J.; Elsts, A.; Mellios, E.; Hilton, G.; Oikonomou, G.; Piechocki, R.; et al. Designing wearable sensing platforms for healthcare in a residential environment. *EAI Endorsed Trans. Pervasive Health Technol.* **2017**, *17*, e1. [CrossRef]

17. CC2650 SimpleLink Multi-Standard 2.4 GHz Ultra-Low Power Wireless MCU. Available online: http://www.ti.com/product/CC2650 (accessed on 1 May 2019).

18. Bluetooth Archived Specifications. Available online: https://www.bluetooth.com/specifications/archived-specifications (accessed on 1 May 2019).

19. IEEE 802.15.4-2015-IEEE Standard for Low-Rate Wireless Networks. Available online: https://standards.ieee.org/standard/802_15_4-2015.html (accessed on 1 May 2019).

20. MX25U6435F. Available online: http://www.macronix.com/Lists/Datasheet/Attachments/7411/MX25U6435F,%201.8V,%2064Mb,%20v1.5.pdf (accessed on 1 May 2019).

21. ADXL362. Available online: https://www.analog.com/media/en/technical-documentation/data-sheets/ADXL362.pdf (accessed on 1 May 2019).

22. Villeneuve, E.; Harwin, W.; Holderbaum, W.; Sherratt, R.S.; White, R. Signal quality and compactness of a dual-accelerometer system for gyro-free human motion analysis. *IEEE Sens. J.* **2016**, *16*, 6261–6269. [CrossRef]

23. LSM6DSO. Available online: https://www.st.com/resource/en/datasheet/lsm6dso.pdf (accessed on 1 May 2019).

24. TPS62746. Available online: http://www.ti.com/lit/gpn/tps62746 (accessed on 1 May 2019).

25. TI RTOS. Available online: http://downloads.ti.com/dsps/dsps_public_sw/sdo_sb/targetcontent/tirtos/index.html (accessed on 1 May 2019).

26. Ghamari, M.; Villeneuve, E.; Soltanpur, C.; Khangosstar, J.; Janko, B.; Sherratt, R.S.; Harwin, W. Detailed examination of a packet collision model for bluetooth low energy advertising mode. *IEEE Access* **2018**, *6*, 46066–46073. [CrossRef]
27. Lee, J.K.; Robinovitch, S.N.; Park, E.J. Inertial sensing-based pre-impact detection of falls involving near-fall scenarios. *IEEE Trans. Neural Syst. Rehabil. Eng.* **2014**, *23*, 258–266. [CrossRef] [PubMed]

Article

A Wearable Closed-Loop Insulin Delivery System Based on Low-Power SoCs

Jesús Berián, Ignacio Bravo *, Alfredo Gardel, José Luis Lázaro and Sergio Hernández

Electronics Department, University of Alcala, Alcala de Henares, Madrid 28805, Spain; jesus@berian.es (J.B.); alfredo.gardel@uah.es (A.G.); josel.lazaro@uah.es (J.L.L.); shsanchez@gmail.com (S.H.)

* Correspondence: ignacio.bravo@uah.es; Tel.: +34-918856580

Received: 9 May 2019; Accepted: 29 May 2019; Published: 31 May 2019

Abstract: The number of patients living with diabetes has increased significantly in recent years due to several factors. Many of these patients are choosing to use insulin pumps for their treatment, artificial systems that administer their insulin and consist of a glucometer and an automatic insulin supply working in an open loop. Currently, only a few closed-loop insulin delivery devices are commercially available. The most widespread systems among patients are what have been called the "Do-It-Yourself Hybrid Closed-Loop systems." These systems require the use of platforms with high computing power. In this paper, we will present a novel wearable system for insulin delivery that reduces the energy and computing consumption of the platform without affecting the computation requirements. Patients' information is obtained from a commercial continuous glucose sensor and a commercial insulin pump operating in a conventional manner. An ad-hoc embedded system will connect with the pump and the sensor to collect the glucose data and process it. That connection is accomplished through a radiofrequency channel that provides a suitable system for the patient. Thus, this system does not require to be connected to any other processor, which increases the overall stability. Using parameters configured by the patient, the control system will make automatic adjustments in the basal insulin infusion thereby bringing the patient's glycaemia to the target set by a doctor's prescription. The results obtained will be satisfactory as long as the configured parameters faithfully match the specific characteristics of the patient. Results from the simulation of 30 virtual patients (10 adolescents, 10 adults, and 10 children), using a python implementation of the FDA-approved (Food and Drug Administration) UVa (University of Virginia)/Padova Simulator and a python implementation of the proposed algorithm, are presented.

Keywords: insulin delivery; wearable system; closed-loop; embedded platform

---

1. Introduction

Diabetes is a condition that affected approximately a 10% of the population during 2015 [1]. Even though there are many subdivisions of diabetes types, like LADA, new-born diabetes, etc., it is still generally accepted the three general categories of type 1, type 2, and gestational. Type 2 diabetes is the most common of the three, but its management is normally only tied to a strict diet combined with exercise. Oral medication is used only when necessary. Only in extreme cases is type 2 treated with insulin. Type 1 diabetes, on the other hand, forces the patient to use insulin as a way to keep the patient's glucose levels in range and reduce the risk of suffering severe short- and long-term complications.

In recent years, we have witnessed, along with new more accurate infusion devices, the appearance of Continuous Glucose Monitoring (CGM), small electronic devices comprised of an electrode, which inserted under the skin, that can measure the glucose levels in tissue fluid. A small transmitter would send the sensor's information to a receiver via radio frequency. Sensors, like the G4 and G5

(Dexcom, San Diego, CA, USA) [2], Enlite (Medtronic, Dublin, Ireland) [3] or Freestyle Libre (Abbott, Chicago, IL, USA) [4], last anywhere from 6 to 15 days and some of them must be calibrated by fingersticks (manual glucose testing) from three to four times per day. Although insulin pumps have been available since the early 60s, their accuracy and functionalities have improved to the point of, for example, being able to dose with precisions of 0.025 units of insulin, integrate data from glucose sensors and glucometers, and keep a history of events for the patient.

Technology has evolved and new possibilities are available for these devices. Developing a closed loop system that reduces the burden faced by people living with diabetes is a priority for medical companies. Most recent devices will not use glucose sensor data to calculate the insulin dosing, even though the pump is receiving this data. Changes to the insulin dosing of a patient could have lethal consequences if mistakes occur. The current systems have no way to cancel the effect of insulin that has been infused, besides the patient consuming carbohydrates. In other words, there is no problem reducing the amount of insulin being dosed as the patient or the system can always increase it later, but once insulin has been dosed the system has no way to remove it from the patient's body. This is the main reason why manufacturers are developing these functions by stages; current devices allow what they call as "suspend-on-low": the insulin pump will suspend the infusion when patient's glycaemia is, or is predicted to be, below a certain security threshold. This safety feature is the first step to obtain a closed loop system. Very recently, a new insulin pump was released to the market that is capable of increasing the insulin dosing depending on the glycaemia. Although this is very critical, the system is preconfigured to achieve a less challenging glucose target while being, at the same time, very conservative and safe.

Every patient is different and so are their needs [5]. That is the reason why a very active community of patients (DIY Artificial Pancreas community/Loop [6]/OpenAPS [7]/Nightscout [8]) have developed their own control platforms and have made them configurable so that, with the necessary knowledge, every patient could adjust it to their own needs. These systems were designed to run on commonly available platforms as Raspberry Pi, Intel Edison or smartphones, typically running Linux or high-end operating systems, and are not designed as low-performance battery powered systems. Other works [9,10] address solutions more complex that do not focus on a wearable approach.

What all of these systems have in common is that they use glucose and insulin dosing information to calculate the actions needed to keep the patient in a defined range (configurable or not by the user). Having a good set of parameters for the patient becomes a critical task that is beyond the scope of this paper.

The rest of the paper is organized as follows: System Proposal will describe the concept of a hybrid closed-loop system, along with some of the problems related to this technology. Section 3, Results, will propose a new system, capable of fulfilling the patient needs as well as reducing consumption and its size compared to the rest of DIY solutions and the results from simulating 30 virtual patients (10 adolescents, 10 adults, and 10 children) using a python implementation of the FDA-approved UVa/Padova Simulator [11] and a python implementation of the proposed algorithm are presented. Section 4, Discussion, will make an analysis of a control loop algorithm and all the possible problems that could occur due to incorrectly set parameters.

2. Materials and Methods

2.1. System Proposal

Our proposed closed-loop system is composed by four different elements: insulin pump, continuous glucose sensor, controller, and data input device (see Figure 1). Some of these elements (or all of them) could be integrated inside the same device.

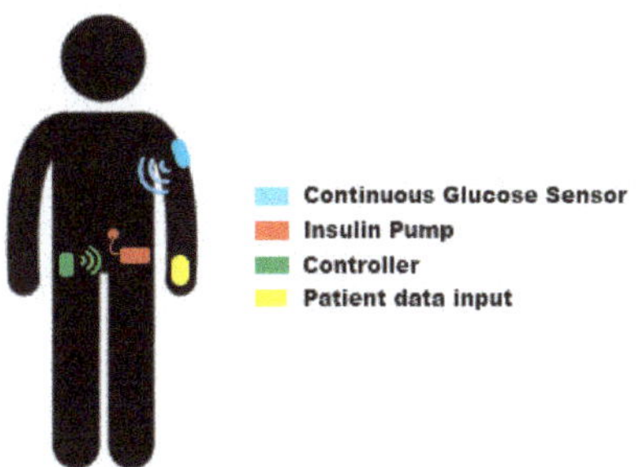

Figure 1. Wearable closed-loop insulin delivery system schema.

Current commercial insulin pumps are designed to work in open loop mode, therefore, they operate in a way that meets the needs for the patients in this situation: insulin pumps allow insulin boluses (equivalent to an insulin injection) and the infusion of basal profiles. A basal profile (commonly named as "basals") defines the different amounts of insulin the patient needs throughout the day to keep his glycaemia stable. As basal needs may change depending on many external factors, pumps typically support more than one basal profile to allow patients to select among them and use the one that best fits their needs in every moment. This basal profile selection is normally done in a type-of-day basis but, if users need to modify their basal infusion for a reduced period of time (from 30 min to some hours) insulin pumps allow "temporary basals." Using a temporary basal the patient can manually increase or decrease the infusion without modifying the programmed basal profile. An insulin pump will be suitable for a closed-loop insulin delivery system if it can be controlled by an external device and supports the change of the infusion configuration: typically setting temporary basals.

The Continuous Glucose Sensor provides the control system with real time information about the patient's glycaemia. The sensor is a module composed by a filament inserted under the patient's skin measuring the glucose concentration and the associated electronics to perform the measurement, and the radio frequency transmission. Current glucose sensors obtain a new reading every 5 min and most of them must be calibrated by the patient in order to get precise enough results. Since the sensor is not measuring directly the glucose concentration present in the blood stream, there is a variable delay between what the glucose sensor is measuring and the real blood glucose concentration. This delay is typically 10/15 min and is one of the reasons why the sensor calibration must be performed when the patient's glycaemia is stable. If the glycaemia is stable both measurements should match when calibrating.

This type of system usually requires input from the patient and that is why they are called "hybrid" closed-loop systems. The main reason for this information need is the delay present in the measurement process and in the action (as shown in Figure 2).

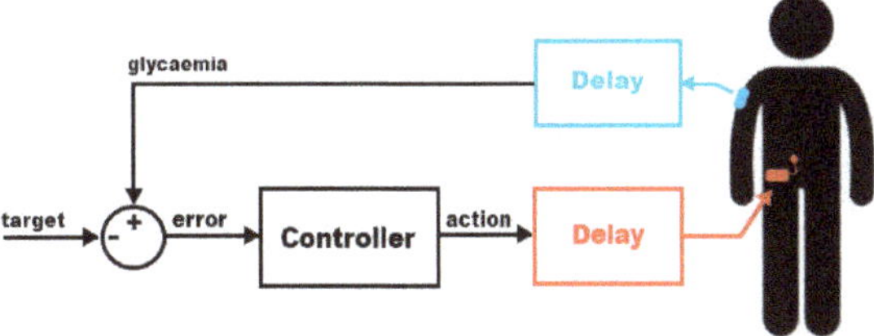

Figure 2. Simplified controller schema with related-to-person behavior and delays.

- Glucose information delay: as explained above, there is a delay of 10/15 min in the glucose measurement process. This reduces the speed in which the controller can act to compensate rapid glycaemia changes.
- Insulin response: the patient will receive insulin via subcutaneous infusion. This process introduces a new delay in the loop system: the effect of the insulin will not be present in the patient's body until it starts being absorbed while in a person without diabetes insulin effect has no delay [12]. Figure 3 compares the relative effects versus time of several types of insulin. As can be seen, the endogenous insulin (the one generated by a healthy pancreas) has an almost immediate effect that ends after one hour of its delivery inside the body. The rest of the insulins are some of the synthetic types a diabetic patient could use. Pump users typically use the fastest insulin available since the long-acting effect can be achieved with programmed basal profiles. Even with the fastest insulin ("Rapid-acting" in Figure 3) the peak effect occurs after 1 h from its infusion and it will be causing an effect up to 8 h.

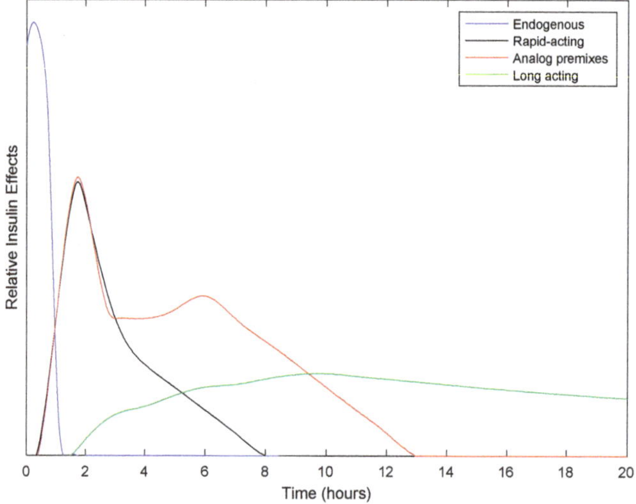

Figure 3. Relative insulin effects versus time.

Although a closed loop system would respond effectively even with these delays inside the loop, they would cause the patient's glycaemia to reach certain levels that are considered dangerous or, at least, not recommended. There are situations in which patients will experience rapid changes in their glycaemia: food intake and exercise. A rich carbohydrate meal will increase rapidly the glycaemia and, by the time the controller is capable to react, the patient could easily reach hyperglycaemia. During exercise, insulin requirements are lower and, if insulin is not reduced soon after, or even before, the exercise the patient could suffer hypoglycaemia. The controller will detect the glycaemia dropping 15 min after the real drop and will reduce the insulin infusion but, as insulin will remain active for many hours (depending on the patient), the effect of the infusion reduction could not be enough. In order to minimize this problem, the controller will receive information from the user to announce meals and exercise before they happen using a device like a smartphone (Figure 4). This way the controller will change the infusion to adapt to the future situation.

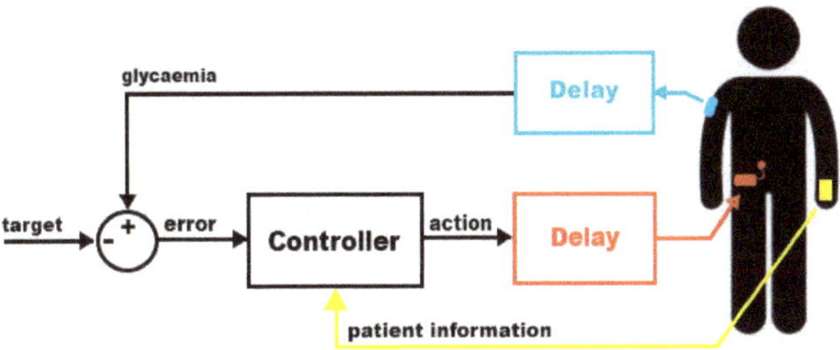

Figure 4. Simplified controller schema with delays and patient information.

The controller is the module responsible for the information analysis and response calculation. In order to be considered as wearable it should meet these requirements:

- Battery with capacity enough to last 24 h in normal operation. Twenty-four hours is the minimum period of time that the controller should be capable of being active between charges so that the patient can recharge it while sleeping.
- Smallest size possible. Patients should be able to store it somewhere in their clothes. To be considered as "wearable," the system must be comfortable, lightweight, and almost unnoticeable by the patient while wearing it.
- Be tolerant to communication failures. The controller will communicate with the glucose sensor and the insulin pump via radio frequency, therefore, the probability of having errors is significant. The controller should be tolerant to errors and should not take actions that, in the event of a loss of communication with the pump, would leave the rest of the system in a dangerous situation.

2.2. Proposed Wearable System for Insulin Delivery

Once the needed specifications for a wearable insulin delivery system are established, in this section, a new proposal that fulfils these requirements will be described. It will be composed by an insulin pump, a glucose sensor, and an ad-hoc controller based on a low power consumption system-on-chip.

The proposed system uses a commercial insulin pump with an integrated glucose sensor (Medtronic Paradigm Veo + Enlite sensor) because its communication protocol via radiofrequency has been documented [13,14] and can be controlled to set temporary basal levels, suspend and resume the infusion and/or command boluses.

As important as having a controllable infusion system is having a calibrated system. Enlite glucose sensors [15] require calibration every 12 h (maximum) and they must be calibrated by the patient following a specific set of rules:

- The glucose sensor will match the reading from the sensor with a glycaemia provided by the patient. This glycaemia value must be obtained with a glucometer following the steps provided by the manufacturer, typically: wash and clean hands before obtaining the blood sample and discard the first drop of blood to avoid external agents from the skin.
- Due to the delay between the glycaemia measured directly in the blood stream and in the interstitial fluids, calibrations must be only performed while the patient's glycaemia is stable. This way the error is reduced as both readings should match after the time equivalent to the delay.

This process is necessary to keep the coherency between measurements in the interstitial fluid (provided by the glucose sensor) and in the blood stream (provided by a glucometer during calibration). The calibration of the sensor becomes really important in a closed-loop system: if the sensor is degraded or not calibrated correctly, the use of inaccurate data could cause a hypoglycaemia or hyperglycaemia to the patient.

Once the patient is using a reliable and controllable insulin infusion system along with a glucose sensor, the next step is to obtain a wearable platform that will enable the patient to close the loop: obtain real time glucose data and patient's parameters and act on the insulin infusion.

Other platforms used for DIY closed-loop insulin delivery systems [6,7] use powerful microprocessors running operating systems as Linux, Android, or iOS. These platforms are not oriented to a low power consumption and they must keep the system running, having other processes in the background that could consume power in an undesired way. This is the reason why these platforms cannot be considered as "wearable." In order to keep these systems running on batteries for at least 24 h, the patient must carry relatively big batteries with the system, making it less portable and, maybe, not reliable for a long-term period of time. When the controller runs on a smartphone there are two additional problems: the patient's phone cannot run out of battery. In that case the patient will open the loop. Other intrinsic problem is the security risk of being hacked. Mobile phone security has to be improved in the next generations of OS for mobile phones. These two problems made this work face the controller's problem from an embedded platform point of view.

In the proposed solution, with low power consumption as a goal, a new custom board was designed. This board is divided into four different parts (diagram in Figure 5):

- Power and battery management: as this will be a battery powered module, some electronics are added to be able to charge the battery and to keep the battery under control.
- Low Power Microcontroller: the control algorithm is coded to run in a Low Power consumption 8051 core. As the controller will receive a new glucose data every 5 min, the controller can sleep the processor during the wait period and reduce the power consumption. Although this microcontroller uses technology from many years ago, its low power consumption makes it the ideal platform for this system.
- Bluetooth Low Energy Communications: this board has a Bluetooth interface to be able to communicate with a smartphone. Using a smartphone, the user can configure the controller and notify external events. It can also be used for remote monitoring purposes like, for instance, Nightscout [8].
- Glucose sensor and Insulin pump Communications: this is the interface used to retrieve information from the sensor and the pump, as well as command actions to the insulin pump.

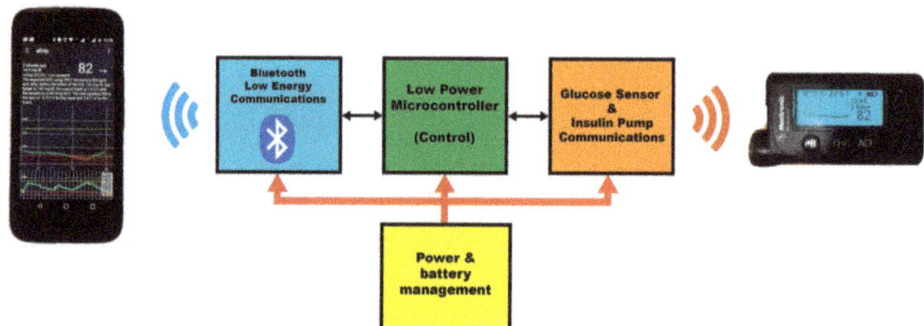

Figure 5. Block diagram of the proposed wearable module.

2.3. Closed-Loop Control

The control loop algorithm will be responsible for the data analysis and actions in the form of temporary basals. Its only purpose will be to command temporary basals to the insulin pump so that the measured glycaemia reaches a certain target value. To do so, the control algorithm was divided into several steps, as shown in Figure 6.

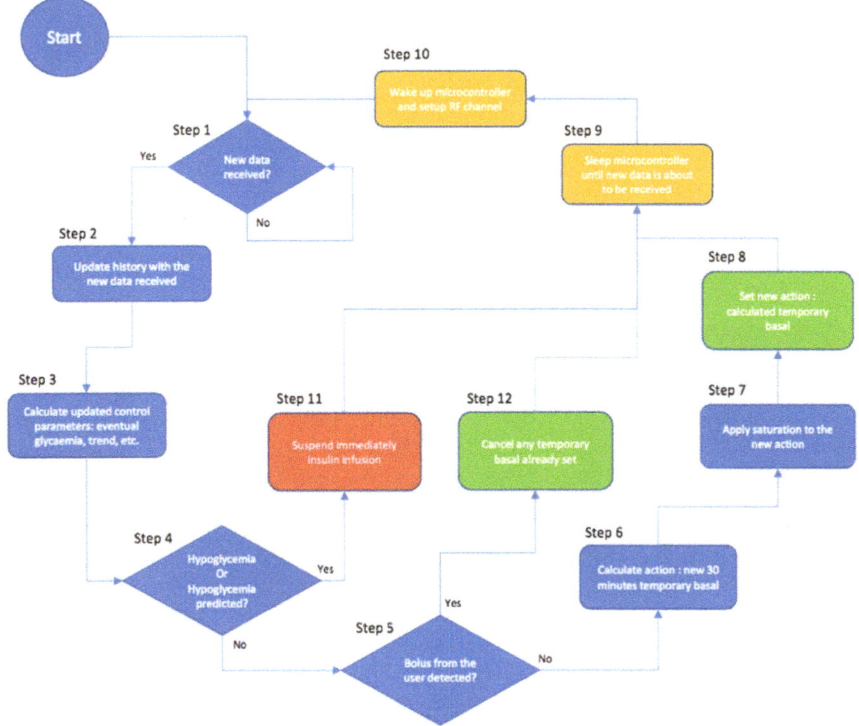

Figure 6. Control loop state diagram.

The control loop takes the following information as input:

- Infused insulin: previously infused insulin will impact the patient's glycaemia even after more than 6 h from its infusion time. This time will vary from patient to patient and also depends on the type of insulin being used.
- Glycaemia: measured by the continuous glucose sensor. Current and past glycaemia values will define the current and eventual status for the patient. The glycaemia is the value the controller will modify to obtain a certain target.

The first thing the control algorithm needs is the updated glycaemia. In step 1 the algorithm will wait until a new glucose value is received from the sensor and, in step 2, the controller will update its history data with the received data.

Once the patient state has been updated, in step 3 the controller will retrieve the pump configuration with some parameters set for and by the patient and, after that, it will calculate some derived parameters needed for the control itself. As every patient needs a different model, it is not feasible to find a universal solution and, therefore, there are some parameters that the patient will need to enter according to his personal characteristics:

- Active Insulin Time: every type of insulin and every patient respond in a different way to the infusion of one unit of insulin. The parameter "Active Insulin Time" makes reference to the time that one unit of insulin keeps affecting the patient's body.
- Basal profile: this profile describes the insulin needs the patients have throughout the day to make their glycaemia stable (in absence of other external factors).

- Sensitivities: Sensitivity or Insulin Sensitivity Factor establishes the relation between one unit of insulin and the glucose concentration that it will reduce in the patient. As happens with the basal profile, the sensitivity changes throughout the day and they are normally configured as a profile.

Once the system has obtained all these parameters and previous information, the control loop will start to operate. The controller will receive a new glycaemia value every 5 min; this allows the system to remain asleep for long periods of time and increase battery life.

As the system has 5 min to do all the calculations needed for the closed loop algorithm, low performance platforms can be used improving power consumption and cost. As a first step, the controller calculates three parameters (eventual glycaemia using only the trend, Insulin-On-Board, and eventual glycaemia) based on all the different inputs:

- Eventual glycaemia using only the trend (Equation (1)): using the history of glucose values sent by the sensor (*sgv or sensor glucose value*, 0 being the latest value received) the controller calculates the trend for the last 'M' values and the glucose value that the patient would reach after 15 min (using T_S and the time in minutes between consecutive glucose values) if the glycaemia continues with that trend. This way, the controller tries to guess the current blood glucose value and reduce the error introduced by the sensor's delay.

$$eGlycTrend = \frac{\sum_{n=0}^{M-1} \Delta sgv[n]}{M} \times \frac{15}{T_S} + sgv[0]. \tag{1}$$

- IOB or Insulin-On-Board for one bolus or dosing (Equation (2)): every bolus or insulin delivered to the patient will affect blood glucose following a certain *"insulin_response"* curve. This curve represents the normalized effect of one unit of the chosen insulin from the infusion moment up to a certain number of hours (active insulin time). This way, the integral of the response curve from the moment chosen for the IOB evaluation (X, measured in hours) to infinity, multiplied by the number of insulin units delivered (B), will be equal to the equivalent amount of insulin that will still have an effect in the patient's body in the future due to that bolus.

$$bIOB(B, X) = B \times \int_{t=X}^{\infty} Insulin\_response_{1U}(t). \tag{2}$$

- IOB or Insulin-On-Board (Equation (3)): using all the previously infused insulin (basal and boluses, taking basals as a series of boluses every 5 min), the controller calculates the amount of insulin that has been infused but has not made an effect yet. That insulin is called Insulin-On-Board and has to be considered when calculating the final glycaemia. IOB will be calculated as the sum of every past infusion IOB (bIOB). The arrival of a new glucose sensor value will trigger the calculation process, so it can be assumed that every bolus or basal insulin can be analyzed as separate boluses with the periodicity of the glucose readings (*Ts*, measured in minutes). Every bolus or infusion will have its corresponding amount of insulin (*Bn*). Variable *n* will serve as discrete index in the equation.

$$IOB = \sum_{n=0}^{\infty} bIOB(B_n, n \times \frac{T_S}{60}). \tag{3}$$

- Eventual glycaemia (Equation (4)): the IOB has an important role: the eventual glycaemia, understood as the glycaemia to which the patient is heading, is calculated by the controller as the eventual glycaemia using only the trend plus the effect of the IOB. The parameter that relates insulin to modifications in the glycaemia is the insulin sensitivity factor (ISF) and is defined as the glucose concentration that one unit of insulin will reduce. The controller will use the insulin sensitivity factor that corresponds to the current time and will multiply it by the IOB previously calculated. This multiplication will give to the controller the expected modification for the glycaemia. Summing up this value to the eventual glycaemia using only the trend the controller obtains the eventual glycaemia.

$$eGlyc = eGlycTrend - IOB \times ISF. \tag{4}$$

At this point, the controller has all the input information and parameters for the patient and all the derived data from them. The controller operation can be divided into four layers evaluated in a sequential manner:

1. Safety Layer (steps 4 and 11 from Figure 6): in this layer, using the current glycaemia ($sgv[0]$), eventual glycaemia ($eGlyc$), and eventually using only the trend glycaemia ($eGlycTrend$) the controller will suspend the insulin infusion in case any of them are below a certain safety threshold ($threshold_{low}$) (see Equation (5)). This same layer will resume the infusion as soon as the hypoglycaemia disappears. In the case of a suspension in this layer, the control loop execution stops until new data is received.

$$suspend\ if \begin{cases} sgv[0] < threshold_{low} \\ \quad\quad or \\ eGlycTrend < threshold_{low} \\ \quad\quad or \\ eGlyc < threshold_{low} \end{cases}. \tag{5}$$

2. Bolus Snooze Layer (steps 5 and 12 from Figure 6): once the controller determines that the patient is not suffering or in risk of a hypoglycaemia, it will check if the patient infused an insulin bolus (step 5). If a bolus was infused, the controller will cancel any temporary basal (step 12) until the peak of effect of the bolus (which depends on the Active Insulin Time). The controller will suppose that the bolus was set to compensate the effect of some food intake and it will not affect the calculation made by the patient for the bolus. After that period of time, the controller will resume the control and compensate if needed

3. Control Layer (step 6 from Figure 6): using the patient's parameters for the current time, the controller calculates the error between the eventual glycaemia and the glycaemia set as target (see Equation (6)). Using the insulin sensitivity factor for that specific time of the day (ISF), the controller is able to calculate the amount of insulin needed to cancel the error (see Equation (7)). That insulin is infused as an addition to the basal needed by the patient, so the controller calculates a new basal rate lasting 30 min (the minimum that the insulin pump permits) that is the result of the basal rate programmed for that period of time (basal) plus the correction insulin distributed through those 30 min (see Equation (8)). If the eventual glycaemia is lower than the target, the resulting temporary basal will be lower than the programmed one for that time.

$$error = eGlyc - target, \tag{6}$$

$$correction = \frac{error}{ISF}, \tag{7}$$

$$tempBasal_{30min} = basal_t + correction \times \frac{60}{30}. \tag{8}$$

4. Action Saturation Layer (steps 7 and 8 from Figure 6): in step 7 the controller limits the actions taken by the control layer. The minimum temporary basal allowed is 0 U/h (Units per hour) and the maximum temporary basal is set as another parameter by the user in the insulin pump. That parameter is called "Max basal." Any temporary basal over that limit will be reduced to the maximum basal allowed. After applying the saturation effect to the calculated temporary basal, it is commanded to the pump in step 8.

As explained before, this control loop has been designed to work on a low-power-consumption embedded platform and, having reached this point in the execution, the control loop will sleep the microcontroller to save energy (step 9 from Figure 6). Just before a new glucose value is sent, the microcontroller will wake up and set up the radio frequency channel to be ready for it (step 10 from Figure 6). The control loop algorithm takes approximately 15 s to calculate and command new actions and, if the glucose sensor sends a new glucose value every 5 min, the microcontroller will sleep for more than 90% of the time.

Although the control algorithm would work properly with some degree of error in the parameters entered by the patient, this controller should be configured as accurately as possible in order to have good results. Some of the effects that the patient could suffer from misconfigured parameters are:

- If the Active Insulin Time does not match the patient's actual one, the expected variations due to the infused insulin will not match the real ones. If the real time is higher and the patient is suffering a hyperglycaemia, the control loop will continue dosing more insulin to correct it without waiting until all the insulin has made effect (possibly causing a hypoglycaemia). On the other hand, if the real time is lower, it will not dose the needed insulin since the controller will wait longer for the insulin to make effect (unrealistically).
- If the basal profile does not match the patient's real one, the corrections will be biased due to that error. For example, in the case of hyperglycaemia, if the programmed basal is lower than the real needs and the correction plus the programmed basal are still lower than the real basal, the controller would never reduce the error and, therefore, take the patient's glycaemia to target. Something similar would happen in the case of hypoglycaemia and a correction plus a programmed basal higher than the real basal, the controller would never reduce the error and would lower the patient's glycaemia (with the risk of causing severe hypoglycaemia).
- If programmed insulin sensitivity factors do not match the patient's real ones the controller will modify the expected response. If the programmed sensitivity is lower than the real one, and depending on the glycaemic error, the controller could infuse too much insulin to correct that error, causing hypoglycaemia. If the sensitivity is higher than the real one, the controller will have a slow response and it will take more time to get to target.

As explained before, some degree of error can be tolerated but, if the error is big enough to cause a non-automatically-solvable situation, an alarm will be sent to the smartphone so that the patient is notified and can take care of the problem.

2.4. Safety

This control algorithm's first goal is to keep the patient safe. Safety features have been implemented in several parts of the system in the form of alarms and action types. As it has been described before, this control platform is based on the fact that the patient is already using a commercial system for glucose sensing and insulin delivery. These systems have their own alarm system being able to report hypoglycemic or hyperglycemic situations directly to the patient. On top of that, our control platform makes use of a third-party software named xDrip [16] to upload data to Nightscout and for data monitoring directly on the phone. xDrip also provides a really powerful alarm system in which the patient can configure the standard alarms (hypo, hyper), rate change alarms or data stalling alarms (for a programmable time period). In the proposed platform, xDrip can also generate alarms coming from the controller. This feature is used to alarm the user from non-mitigatable problems like a precited hypoglycemia that cannot be solved just by reducing the future amount of insulin being delivered, requiring the patient interaction or a persistent hyperglycemia that the system fails to lower, meaning that the patient may need to review the infusion set or the correct state of the insulin.

Safety was also taken into account in the way insulin is controlled. There are two main ways of controlling infusion:

- Micro-bolusing: the insulin pump only injects insulin when the controller commands it in the form of small boluses.
- Temporary basals: the insulin pump has a programmed basal rate profile that should cover the patient's basal needs during the day and the controller will issue temporary basal rates to change the insulin infusion when needed.

If the control algorithm resides inside the insulin pump any of those methods can be understood as safe but, as we have to rely on the radiofrequency communications that could or could not be operational, micro-bolusing is not a safe option: to be able to use micro-bolusing the patient must disable any basal rate infusion since there would be no other way to reduce the amount of insulin the patient receives below that basal rate (there would be no way to suspend infusion). Disabling the basal rate is potentially dangerous since any problem with the controller or the communications would leave the patient without insulin and that situation could lead the patient to suffer from DKA (Diabetic Ketoacidosis). Due to this reason, this controller makes use of temporary basal rates to control infusion. The patient has a valid basal rate profile configured and the controller will issue 30 min temporary basal rates to add or reduce insulin. In the case of a controller or communications failure, after those 30 min, the insulin pump will come back to its normal behavior and will infuse the programmed basal rate.

One more problem mitigated with the use of temporary basal rates is that, in the case of an uncontrolled loop execution (execution error event), one temporary basal will directly cancel the previous one while, in the case of micro-boluses, the system could potentially command many boluses in a row stacking their effects (and possibly causing a severe hypoglycemic event).

With the use of temporary basal rates the controller has another safety feature: the insulin pump will only accept temporary basal rates that are higher than 0 U/h and lower or equal than a maximum rate that can be configured in the pump, this way constraining the possible actions that can be taken by the controller.

2.5. Security

One of the main concerns about using DIY systems [6,7] with outdated Medtronic pumps, apart from glycemic safety, is the lack of security in the communications channel between the controller and the insulin pump. Communications are not encrypted in any way and, while this fact makes the insulin pump easily accessible to the controller, it also makes it vulnerable to attackers who could try to remote control the pump and, therefore, be a potential risk for the user's safety.

The controller will query the pump every 5 min and, by doing so, the pump ID is broadcasted several times with every data exchange. Every message will contain the pump ID so that it can be identified by the receiver and any other system waiting for those messages. These periods of time are critical since the pump ID is exposed and any attacker could potentially listen to it and use it later to send dangerous commands to the pump.

Medtronic has created a new communications protocol to address this problem encrypting the payload of the messages but, as this system makes use of the old model ones, an intrusion detection and jamming procedure was added to the controller.

The controller is always listening, looking for messages from/to the pump. If a message is received and it is not an answer to a command sent by the controller, an intrusion alarm will be sent to the smartphone to alert the user. At the same time, the jamming procedure is started: the controller will continuously send "Suspend" messages to the pump in an attempt to cancel any command sent to the pump while giving time to the user to take any action needed to protect himself from this situation. This continuous transmission will corrupt any message sent by the attacker, blocking the communications channel for some minutes.

This protection can be disabled to allow data download from the pump when the user needs it (for instance during a medical appointment).

3. Results

A new hardware platform (see Figure 7) was developed, following the previously mentioned requirements, to create a new automatic insulin delivery system based on the patient's glycaemia and his specific parameters. Three system-on-chip ICs were included to support communications with Medtronic devices (Texas Instruments CC1110 [17]), Bluetooth Low Energy devices in Central Role mode (Texas Instruments CC2540 [18]) and, although it is not necessary right now, other 2.4 GHz devices for future developments (Texas Instruments CC2510 [19]). All three system-on-chips apart from the RF transceiver incorporate an 8051 processor and were used to implement, among other functionalities, the control loop algorithm (see Figures 7 and 8). All of them can enter really low power modes in which they reduce their consumption down to 0.3 µA (PM3 mode).

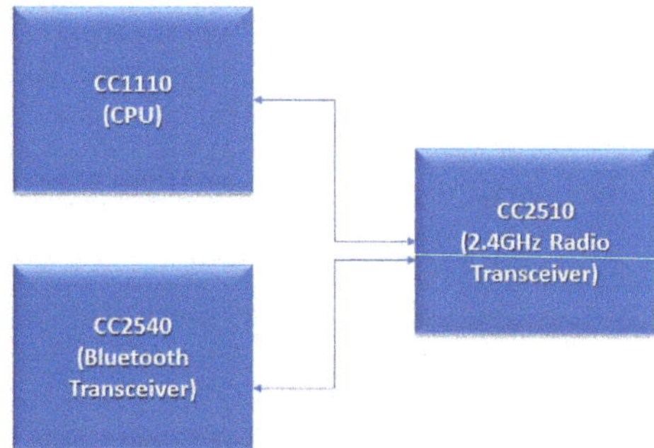

Figure 7. System-On-Chip Interconnection block diagram.

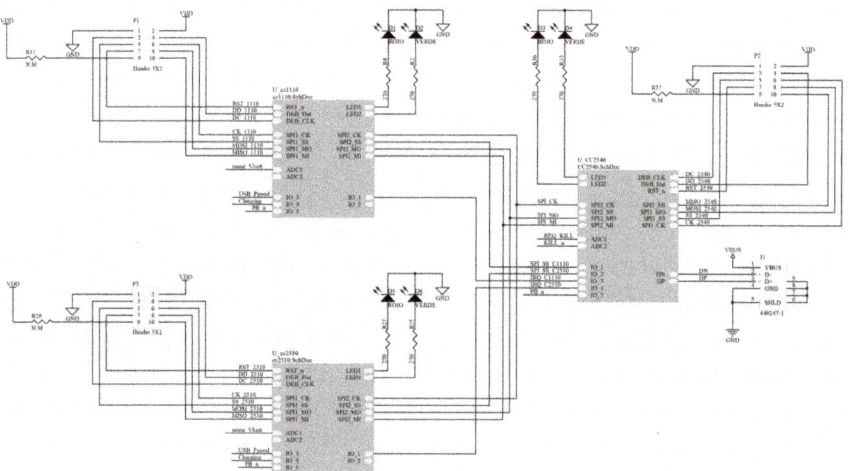

Figure 8. Designed platform—SoC interconnection.

As this module has to be wearable, a 680 mAh Lithium-Polymer battery with its regulation electronics were added (see U1 and U3 in Figure 9). This battery provides more than 24 h of operation enabling the patient to safely wear it during the day and recharge it during the night while still operating.

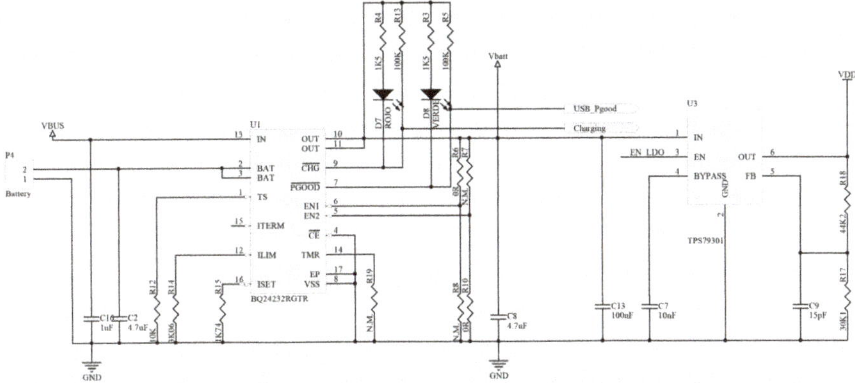

Figure 9. Designed platform—battery and power regulation.

In addition, this new platform was integrated into the Nightscout platform [8] for remote monitoring and with xDrip [16] as a mobile platform to control it and upload information to the cloud. Therefore, this platform (shown in Figure 10) can be connected with cloud solutions to share and analyze information.

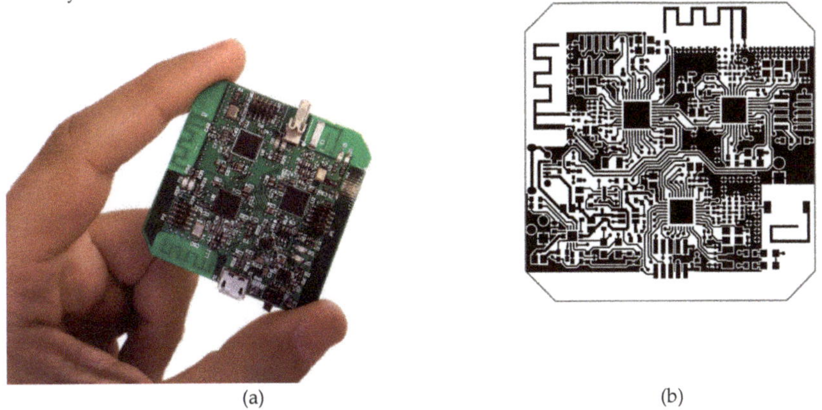

(a)   (b)

Figure 10. Hardware platform developed for the closed-loop insulin delivery system. (**a**) Device picture, (**b**) PCB. Top layer.

Using Nightscout, the user can control the performance of the loop and receive alarms when needed. Nightscout, as a platform, is a perfect way to adjust parameters as it enables the patient to log the information for meals, boluses, exercise, etc., and use this information for later analysis. Endocrinologists can benefit from this tool to make better adjustments as more information can be easily analyzed.

Figure 11 shows the typical Nightscout webpage a patient would see in normal operation. The big green number on the top right corner (item number 1) shows the current glycaemia and an arrow showing the trend. The dotted line (item number 7) shows the patient's glycaemia in real time. In this case, the target range (green) is set to be 90–140 mg/dL. The bottom line (item number 8) shows two days of data while the middle one can show a configurable period of time (last 3, 6, 12 or 24 h) using selector number 6. The blue graph on top of the glycaemia line (item number 9) shows how the insulin infusion has been performed. Other parameters that can be monitored are:

- Current time (item number 3)

- IOB: current insulin on board
- Numeric trend (item number 2)
- COB: carbohydrates on board (if the user enter meal information)
- CAGE: Cannula Age. As a reminder for the patient to change the cannula when needed.
- SAGE: Sensor Age. As a way to monitor the time the sensor has been placed in the patient's body.
- IAGE: Insulin Age. Show the time since the insulin reservoir was filled with insulin.
- BASAL: current basal or temporary basal set.
- PUMP: current insulin pump status (insulin left in the reservoir and battery voltage)
- Time since last reading and since last action from the controller (item number 4).
- Battery level for the smartphone uploading the data to the cloud (item number 5).

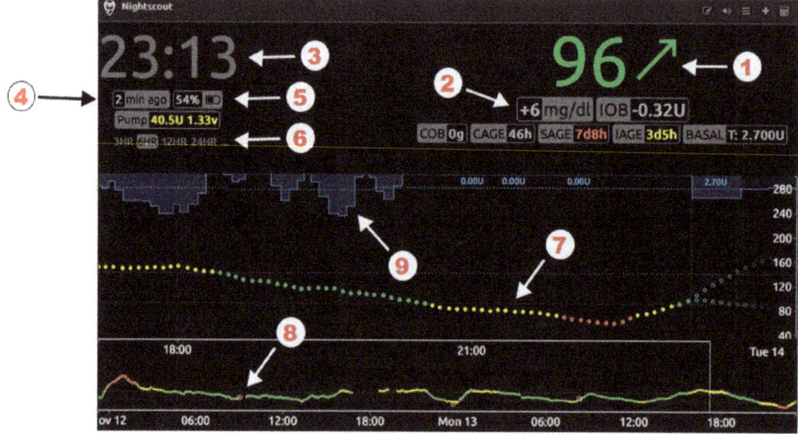

Figure 11. Nightscout.

The proposed control algorithm will obtain good results as long as the user configures the parameters that match his/her needs (active insulin time, basal profile, and insulin sensibility factors) [20]. To test this control algorithm, an open-source simulator named simglucose [11] was used. This simulator is a python implementation of the FDA-approved UVa/Padova Simulator [21] for research purpose only that includes 30 virtual patients: 10 adolescents, 10 adults, and 10 children.

The patients' configuration for the simulated insulin pump were set following standard methods and, as the parameters that vary from patient to patient are included in this configuration and their calculation is out of the scope of this article, it is assumed that results should not significantly vary among other possible patients. The algorithm should take the same decisions that the patient would normally take if he/she would be doing it manually every 5 min.

Three batches of simulations were run: one of them in open-loop mode without any intervention from the patient, one in open-loop mode but simulating the correction action from the patient administering correction boluses 2 h after every meal, and the other in closed-loop mode. All simulations used the same set of parameters and, although they differ from patient to patient, they were not modified from simulation to simulation. The following meal scenario was used:

- 07:00 -> 40 grams
- 13:00 -> 20 grams
- 17:00 -> 5 grams
- 20:00 -> 20 grams

To evaluate results coming from these simulations, the "Time In Range" (TIR) parameter was used. "Time in Range" refers to the percentage of time a patient spends within his/her target glucose range, typically from 70 to 180 mg/dL. The most time the patient stays inside this range, the less likely he/she is to suffer from long-term complications in the future.

In open-loop mode without interaction from the patient, as can been seen in Figures 12 and 13, most patients suffer from hyperglycemia more than 40% of the time. Figure 13 shows how glucose levels start rising with time and correction boluses would be necessary to put the patient back in range (70–180 mg/dL). Meal absorption is shown in terms of CHO (grams). CHO stands for Carbohydrates (Carbon Hydrogen Oxygen) and, in these simulations, the rate of absorption was set to 5 g per minute. Since the simulation step is 5 min, CHO is equivalent to the grams of the meal divided by five. This simulation obtained the following average TIR values: general 40.52%, adolescent 52.45%, adult 37.71%, child 31.41%.

In an attempt to simulate part of what a patient would do to correct hyperglycemias, a simulation in open-loop mode but allowing correction boluses was executed. These correction boluses occur 2 h after every meal only if hyperglycemia exists. As can be seen in Figures 14 and 15 results improved, reaching a global average TIR percentage of 65.75%. The average TIRs per groups were: adolescent 77.78%, adult 67.23%, child 52.25%.

In closed-loop mode, simulating the same patients with the same meal scenario, TIR improves considerably as can be seen in Figure 16. Figure 17 shows how glucose levels are kept in range most of the time, and some patients suffer from small post-prandial hyperglycemic episodes. These episodes could be minimized applying pre-bolusing techniques (dosing insulin some time prior to the meal) but this simulator does not support it directly. This simulation obtained the following average TIR values: general 93.69%, adolescent 93.11%, adult 97.99%, child 89.96%. One really important thing to point out is that using a closed-loop system not only improves TIR but also reduces the burden the patient suffers. The system operates autonomously, and the patient does not need to spend so much time trying to keep his/her glucose levels in range.

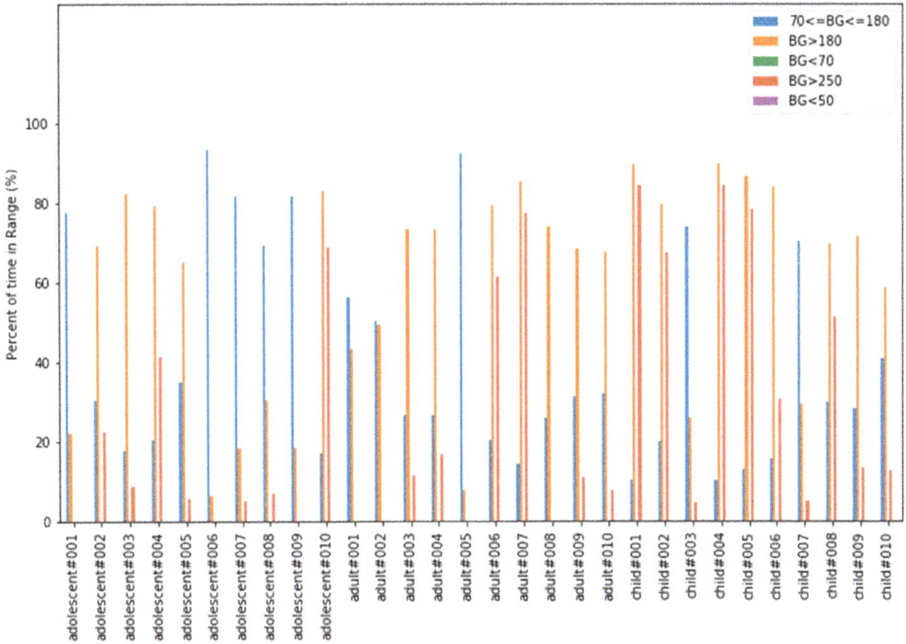

Figure 12. Open-loop simulation–Time In Range.

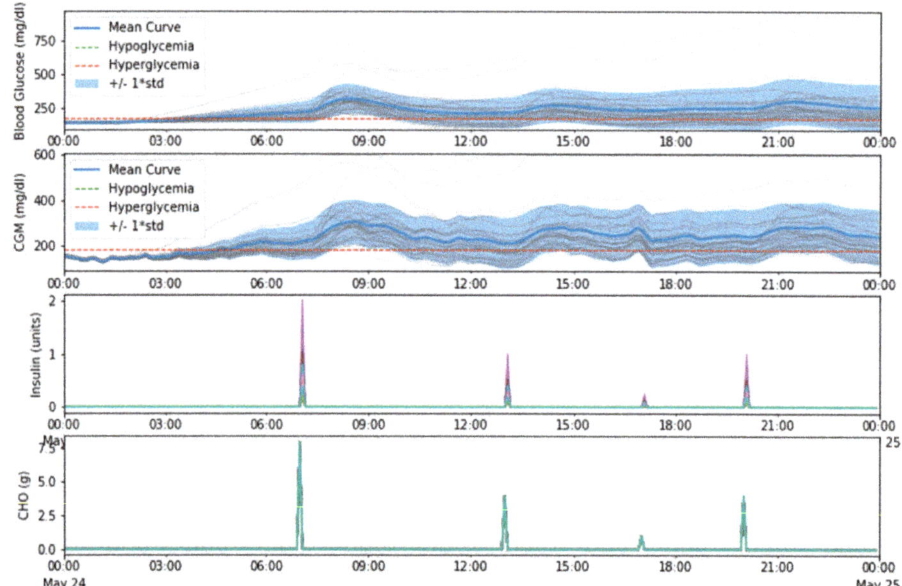

Figure 13. Open-loop simulation—mean results.

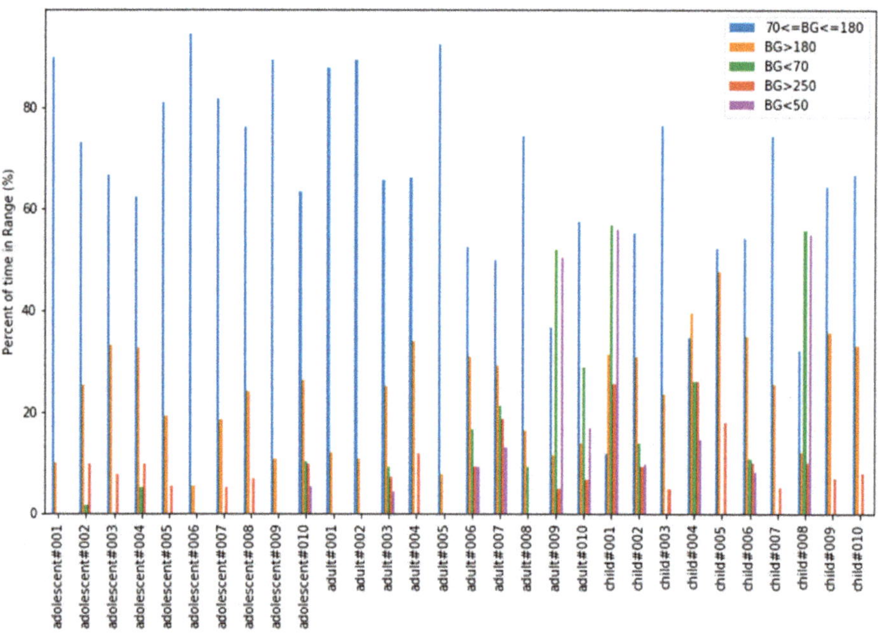

Figure 14. Open-loop with corrections—Time In Range.

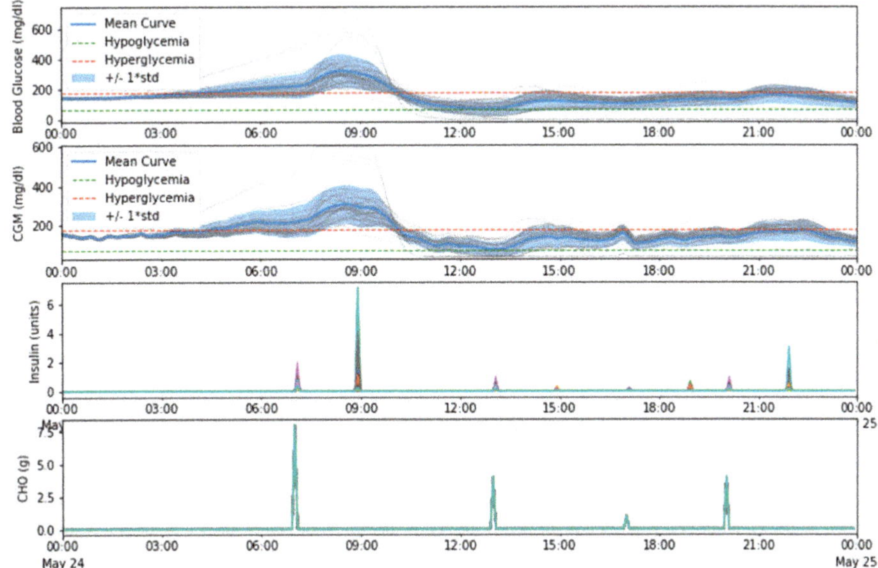

Figure 15. Open-loop with corrections—mean values.

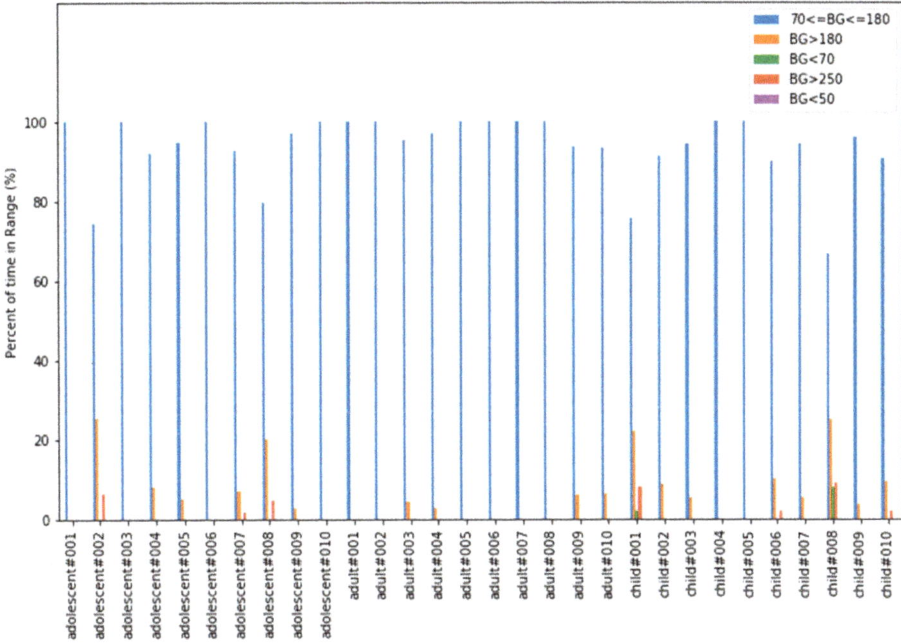

Figure 16. Closed-loop simulation—Time In Range.

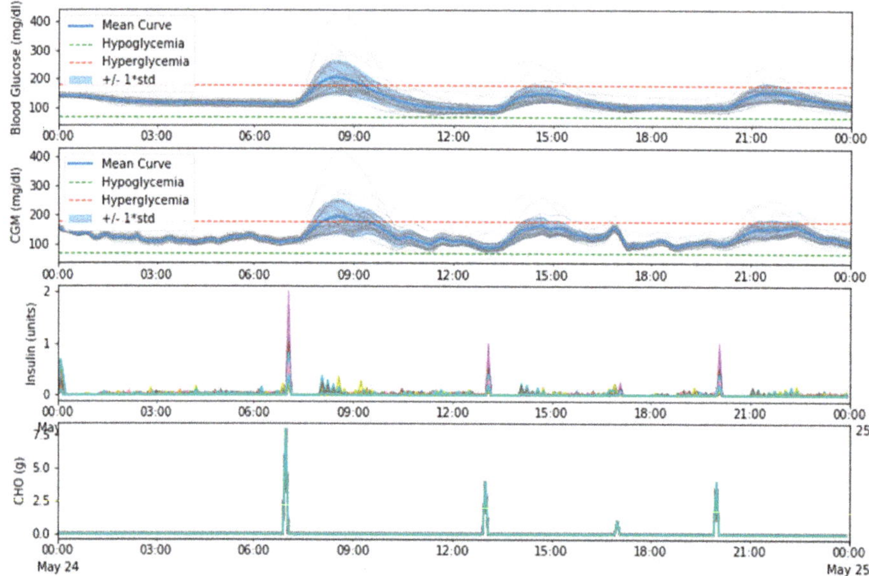

Figure 17. Closed-loop simulation—mean results.

4. Discussion

The typical recommendation when using a controller like this one is to stay in open loop mode for at least two weeks fine tuning these parameters before activating the closed-loop control. Unfortunately, setting these parameters correctly is not an easy task. Even with a controller perfectly configured, there are situations that will temporarily modify these parameters and that are not under the patient's control: incorrectly estimated food intakes, illness, exercise, stress, hormonal disorders, and many others.

As these parameters can vary easily (even for one single patient) it seems reasonable to think that new models are needed so that these parameters can be estimated and that, in this way, any controller could do a better job. This is a good opportunity for the scientific community to use open platforms and obtain better control algorithms.

This controller offers a real solution that allows the patient to keep his glycaemia under control while being, at the same time, almost transparent and unnoticeable. Therefore, the proposal described in this work provides a new approach to improve the life of patients with this disease taking advantage of a glucose sensor, an ad-hoc processing unit, and a set of wireless channels of communication.

Author Contributions: Conceptualization, J.B. and S.H.; Methodology, I.B., A.G. and J.L.L.; Software, J.B. and S.H.; Validation, J.B. and I.B.; Investigation, J.B.; Writing—original draft preparation, J.B.; Writing—review and editing, I.B, A.G. and J.L.L.

Funding: This research received no external funding.

Acknowledgments: This system was created using other modules as a base: Nightscout was used for remote monitoring and visual control, xDrip Android App was modified to support this new platform and to upload glucose and control information in Nightscout. The insulin pump and the glucose sensor used were developed by Medtronic.

Conflicts of Interest: The authors declare no conflict of interest. The founding sponsors had no role in the design of the study; in the collection, analyses, or interpretation of data; in the writing of the manuscript, and in the decision to publish the results.

References

1. American Diabetes Association. Available online: http://www.diabetes.org/diabetes-basics/statistics (accessed on 9 December 2018).
2. Dexcom. Available online: www.dexcom.com (accessed on 9 December 2018).
3. Continuous Glucose Monitoring—Medtronic. Available online: https://www.medtronicdiabetes.com/treatments/continuous-glucose-monitoring (accessed on 9 December 2018).
4. Abbott Freestyle Libre. Flash Glucose Monitoring System. Available online: https://www.freestylelibre.us (accessed on 9 December 2018).
5. Lee, H.; Buckingham, B.A.; Wilson, D.M.; Bequette, B.W. A Closed-Loop Artificial Pancreas Using Model Predictive Control and a Sliding Meal Size Estimator. *JDST* **2009**, *3*, 1082–1090. [CrossRef] [PubMed]
6. Github repository for Loop. Available online: https://github.com/LoopKit/Loop (accessed on 9 December 2018).
7. The Open Artificial Pancreas System #OpenAPS. Available online: https://openaps.org (accessed on 9 December 2018).
8. Nightscout (CGM in the Cloud). Available online: http://www.nightscout.info (accessed on 9 December 2018).
9. Doyle, F.J.; Huyett, L.M.; Lee, J.B.; Zisser, H.C.; Dassau, E. Closed-Loop Artificial Pancreas Systems: Engineering the Algorithms. *Diabetes Care* **2014**, *37*, 1191–1197. [CrossRef] [PubMed]
10. Kudva, Y.C.; Carter, R.E.; Cobelli, C.; Basu, R.; Basu, A. Closed-Loop Artificial Pancreas Systems: Physiological Input to Enhance Next-Generation Devices. *Diabetes Care* **2014**, *37*, 1184–1190. [CrossRef] [PubMed]
11. Xie, J. Simglucose v0.2.1 (2018). Available online: https://github.com/jxx123/simglucose (accessed on 9 December 2018).
12. Freeman, J.S. Insulin Analog Therapy: Improving the Match with Physiologic Insulin Secretion Devices. *Clin. Prac.* **2009**, *109*, 26–36.
13. Medtronic Veo RF Transceiver. Github repository. Available online: https://github.com/jberian/mmcommander (accessed on 9 December 2018).
14. Medtronic Veo Commands. Github repository. Available online: https://github.com/bewest/decoding-carelink/tree/master/decocare (accessed on 9 December 2018).
15. Medtronic Enlite Sensors. Available online: https://www.medtronicdiabetes.com/customer-support/sensors-and-transmitters-support (accessed on 9 December 2018).
16. xDrip—Using your Dexcom G4 with Android. Available online: http://stephenblackwasalreadytaken.github.io/xDrip (accessed on 9 December 2018).
17. Texas Instruments Low-Power SoC CC111x Datasheet. Available online: http://www.ti.com/lit/ds/symlink/cc1110-cc1111.pdf (accessed on 9 December 2018).
18. Texas Instruments Low-Power SoC CC2540 Datasheet. Available online: http://www.ti.com/lit/ds/symlink/cc2540.pdf (accessed on 9 December 2018).
19. Texas Instruments Low-Power SoC CC2510 Datasheet. Available online: http://www.ti.com/lit/ds/symlink/cc2510.pdf (accessed on 9 December 2018).
20. Kovatchev, B.P. Diabetes Technology: Markers, Monitoring, Assessment, and Control of Blood Glucose Fluctuations in Diabetes. *Scientifica* **2012**, 283821. [CrossRef] [PubMed]
21. Man, C.D.; Micheletto, F.; Lv, D.; Breton, M.; Kovatchev, B.; Cobelli, C. The UVA/PADOVA Type 1 Diabetes Simulator: New Features. *JDST* **2014**, *8*, 26–34. [CrossRef] [PubMed]

 *electronics*

Article

On the Design of Low-Cost IoT Sensor Node for e-Health Environments

Nikos Petrellis [1],*, Michael Birbas [2] and Fotios Gioulekas [3]

[1] Department of Computer Science and Engineering, ATEI of Thessaly, 41110 Larissa, Greece
[2] Department of Electrical and Computer Engineering, University of Patras, Campus of Rion, 26504 Patras, Greece; mbirbas@ece.upatras.gr
[3] Department of Informatics, Aristotle University of Thessaloniki, 54124 Thessaloniki, Greece; gioulekas@csd.auth.gr
* Correspondence: npetrellis@teilar.gr; Tel.: +30-2410-684542

Received: 31 December 2018; Accepted: 29 January 2019; Published: 2 February 2019

Abstract: The proliferation of Internet of Things (IoT) devices for patient monitoring has gained much attention in clinical care performance, proficient chronic disease management, and home caregiving. This work presents the design of efficient medical IoT sensor nodes (SNs) in terms of low-cost, low power-consumption, and increased data accuracy based on open-source platforms. The method utilizes a Sensor Controller (SC) within the IoT SN, which is capable of performing medical checks supporting a broad coverage of medical uses. A communication protocol has been developed for data and command exchange among SC, local gateways, and physicians' or patients' mobile devices (tablets, smart phones). The SC supports moving average window (MAW) and principle component analysis (PCA) filtering algorithms to capture data from the attached low-cost body sensors of different sampling profiles. Significant extensions in SN's portability is achieved through energy consumption minimization based on the idle time gaps between sensors' activations. SN's components are either deactivated or set to low activity operation during these idle intervals. A medical case study is presented and the evaluated results show that the proposed SN can be incorporated into e-health platforms since it achieves comparable accuracy to its certified and high-cost commercial counterparts.

Keywords: e-health environment; sensor applications; patient monitoring; medical IoT sensor nodes; low-power; accuracy; IoT platforms

1. Introduction

E-health monitor infrastructures can improve the quality of life in elderly people [1–3] while the advent of wearable sensors could also help in chronic disease management [4]. Typical medical checks or vital signals monitoring are performed by the aforementioned healthcare systems and data are utilized by physicians to monitor health conditions. The combination of low-cost devices and advanced information systems can reduce the cost of daily life monitoring while preventing unnecessary hospital admissions. Although high precision is the main target that has to be accomplished in the medical checks performed by e-health monitoring systems, low cost can encourage more users to adopt them. Additionally, the Internet of Things (IoT) technology is utilized to reduce costs in IoT-based healthcare services towards the support of a ubiquitous computing environment with medical data processing, monitoring, and prediction [5]. The IoT infrastructure is either based on expensive commercial devices or open-source technologies of low implementation cost. Patients of severe heart diseases were monitored by an e-health platform [6] that utilized commercial monitor devices and sensors [7,8] to perform Electrocardiograms (ECGs) in real-time. In this platform, Wi-Fi or Global System for Mobile communications (GSM) telecom protocols were employed to send sensor information to

IoT infrastructure. The work in [9] proposes a similar monitoring architecture utilizing on-the-shelf sensor devices of high-cost while researchers in [10] focus on the implementation of a middleware to interface wireless protocols with the sensor devices and the rest of the IoT platform so as to transform raw-data to patient information without taking into account low-energy consumption or low-cost sensors. The work in [11] proposed a patient monitoring environment based on wearables by utilizing high-precision and low-power sensors. The authors in [12] developed a body area network of sensors that recorded vital signals stored in a database from where physicians could check the health status of individuals. The proposed system was based on Arduino and Raspberry PI as combined in the e-Health Sensor Shield [13]. However, no information was given about the features that are vital for this kind of an infrastructure (e.g., data communication mechanism, energy consumption, and accuracy).

The same pair of popular computing platforms (Arduino and Raspberry PI) are also used in [14] for monitoring the breeding behavior of vultures in their nests. A tiny version of Arduino called Miniduino connected to a number of sensors (temperature, accelerometers, etc.) is camouflaged as an egg. The data gathered are sent to a "data-relay" implemented with Raspberry PI through Bluetooth. The data relay, connected to the Internet, stores the sensor values in a data cloud. Although this application differs from health monitoring, the system architecture and the challenges faced are similar (power consumption, communication method between sensor controller and data relay, etc.). The in-home e-health monitoring platform that was proposed in [15], consists of an expandable WSN, which sends its sensor samples to an in-premises gateway (GW) through communication protocols that follow the Serial Port Profile (SPP) [16]. The gathered data are transmitted to a service based system that is responsible for database transactions, data processing, and medical assessment (e.g., vital signals, alerts). Heterogeneous sensors with different real time operating systems (tinyOS, Android, etc.) and communication interfaces are connected to the Gateway. However, this framework does not provide details for the connection of analog sensors.

The work in [17] designed a power-over-Ethernet GW to perform proper data filtering and streaming so as to send individuals' vital signals to monitor devices. The MIThril system [18] uses body and ambient wearables, which are incorporated into a WSN and restricted to only short-term tele-monitoring. On the other hand, there is a plethora of proprietary IoT platforms like e-Shimmer Health BSN [19] or Simband health sensor [20] from Samsung, with proprietary components that don't provide open-source hardware (HW) features. The cost of such an individual commercial sensor module may be many (2, 3, or more) times higher than the overall cost of the controller and the set of sensors used in our platform.

Approaches towards the exploitation of open hardware platforms for the implementation of IoT SNs and gateways have been proposed recently in the literature [21–28]. To this end, medical sensors are incorporated and plugged into the aforementioned platforms for monitoring patient vital signals while maintaining low-cost equipment and facilities.

Within the context discussed above, current work focuses on the development of an IoT Sensor Node which incorporates low-cost sensors to monitor elderly, disabled, or even healthy persons in non-life threatening situations. Our targets are:

- First, to design and implement a low-power communication protocol to efficiently deliver sensor data from the IoT SN to the medical staff that monitors the patient.
- Second, to improve the quality of the received sensor data from low-cost sensor devices by enhancing their accuracy through the usage of filtering algorithms.
- Third, to decrease energy-consumption and increase battery life-time ensuring portability.

Specifically, the developed system can be used to merely monitor a person's habits like the amount of time being spent in bed, or it can be used to perform trivial medical tests like body temperature, blood pressure measurement, etc., or it can even support more advanced medical tests like ECG. Although the use of the developed computing platform can be scaled to multiple patients monitoring as the case is in a hospital environment, its main use is targeted to the monitoring of individual persons in their own environment. The low-cost e-Health kit of Libelium/Cooking Hacks, mentioned before, was also utilized by the authors towards the development of e-health monitoring systems [29,30]. The architecture of this platform is introduced in [29]. The methods employed for low power consumption and higher sensor accuracy are briefly described in [31–33]. Libelium discontinued the employed version (v2.0) of the e-Health platform and its successor called MySignals supports additional sensors with higher precision. However, the cost of the new MySignals platform is comparable to the cost of similar commercial products. Since our target is to keep the cost low we have developed methods that improve sensor accuracy and minimize power consumption using the older version v2.0. In this regard, several architectural features of the v2.0 version can be integrated in a new e-health system that could be offered as a low-cost platform that achieves though, high precision measurements flexibility and portability with low energy consumption. Specifically, in contrast to prior developments [29–33], we have significantly improved the platform's functionalities by performing suitable procedures and hardware/software expansions. In this context, it was achieved to support a diversity of medical scenarios using an advanced control and communication mechanism (between the SN and the GW) in order to plan the proper time intervals where sensors' sampling is performed based on user requirements. Moreover, we have activated low energy consumption methods that utilize the time gaps (idle-intervals) between successive sensor measurements in the sense that these are translated to appropriate "sleep mode" operation intervals. During these intervals, some of the SN's processes are deactivated periodically so as to expand portability by prolonging its battery life-time. Furthermore, it was possible to significantly enhance the precision of specific sensors, making them in many cases competitive with certified medical devices as shown in our comparison study. Overall, by providing the developed platform with the aforementioned value added characteristics it is aimed to come up with a low-cost and advanced e-health sensor node suitable to accommodate a plethora of medical scenarios and checks.

The rest of the paper is organized as follows: Section 2 describes the proposed e-health infrastructure while focusing on the IoT SN's features and its implementation. Furthermore, the work regarding enhancement of the sensors precision is also presented. This section analyzes the power dissipation issues whereas the applied techniques for power consumption reduction are introduced. Furthermore, the implemented communication protocol between the SC and the Local GW is also presented. Section 3 describes a demonstrative medical test case to show the efficiency of the proposed methods and procedures. Finally, we demonstrate and discuss the experimental evaluation results of the sensors of the developed IoT Sensor Node, including a comparison with commercial certified sensors where possible.

2. Materials and Methods

2.1. IoT Sensor Node Design

The IoT Sensor Node, developed in this work, is applicable to an e-health monitoring framework like the one in Figure 1, where a patient is monitored either in a hospital or a rehabilitation center. Orders can be given by a supervising physician for specific medical tests that can be performed by the patient himself or with nurse assistance, while the results are uploaded to a cloud or the Hospital's Information System (HIS). Qualified staff can examine the results of the medical tests in order to define appropriate treatment or schedule additional tests. The IoT Sensor Node comprises an Arduino UNO Microcontroller and a Raspberry PI platform [13]. A set of medical sensors with either a simple analog or smart digital interface can be attached to the monitored person and they are all connected to the portable IoT Sensor Node (through the sensor coupler). A low power circuitry has been designed and incorporated as power control mechanism. The IoT Sensor Node communicates through the connectivity layer to a Local IoT Gateway which is installed on the premises. The Gateway gives comprehensive instructions to the smart phone or the tablet of the patient under supervision.

The Gateway is also used to forward the measured sensor values to a database or cloud which is accessed by the supervising physician or nurse [5]. In the reverse direction, the Gateway accepts a scenario of medical tests that should be performed, translating this scenario into commands that are executed by the Sensor Controller (SC). The SC implements a protocol that is executed on the Raspberry PI processor and supports the communication with the Local Gateway using a wireless infrastructure (e.g., Wi-Fi, Bluetooth). The SC collects data from sensors and transmits them to the Gateway. Slight tasks like alerts regarding temperature, humidity conditions, patient behavior etc. can be directly retrieved via the GW, whereas, of course, the supervising doctor makes the significant conclusions by taking into account the stored medical values. At this point, it should be mentioned that although, as aforementioned, the developed system mainly aims to remote monitor individuals in their own environments, its use for a higher number of persons is not restricted as long as the supervising doctors are able to monitor all the sensor values stored in the cloud (or HIS) and the necessary privacy and security levels are preserved. The local gateway usually transmits its data to the cloud or HIS by employing the Message Queuing Telemetry Transport mechanism (MQTT protocol). The MQTT protocol utilizes publish-subscribe communication over the Transmission Control Protocol (TCP) layer (employing shorter header than typical TCP), and is used for Machine-to-Machine IoT applications, due to its flexibility and scalability [34].

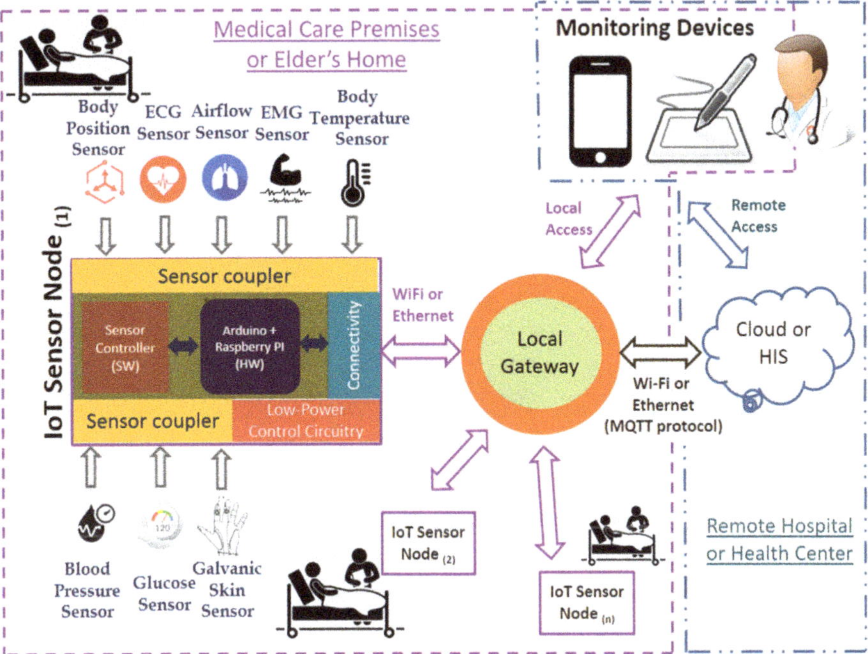

Figure 1. Architecture of the IoT e-Health environment. The IoT SN consists of the Body Sensors, the Sensor couplers, the circuitry system for low power control, the SC as SOFTWARE program that operates on the open source HW of Arduino and Raspberry PI and the connectivity infrastructure (Wi-Fi and Ethernet card). The sensor data are collected by an in-premises Gateway, which in turns sends them to the cloud platform based on the MQTT protocol. Each IoT SN is connected to the relevant individual under monitoring. Smartphones or Tablets can be used to monitor the vital signals of the individual either by connecting to the local gateway or to the cloud platform (or HIS).

Regarding security, we utilize the encryption as provided by the wireless communication method employed like IEEE 802.11, provided that all the persons that can connect to the local access point are authenticated. The File Transfer Protocol (FTP)/Hypertext Transfer Protocol (HTTP) server accessibility is also subject to authentication when sampling commands have to be downloaded or sensor data to be uploaded to the gateway. Authentication is also mandatory for access of the Real Time Operating System (RTOS) that runs on the IoT SN for debugging purposes. If more advanced security methods are required, libraries like OpenSSL that are compatible with many RTOS can be used. The additional encryption methods supported by OpenSSL can hide sensitive medical data exchanged between the IoT SN and the Gateway from other users that are connected to the Local Access Point. The security and access rights of the data stored in the remote cloud can be guaranteed by advanced private and public key distribution methods, but their study is out of the scope of this paper.

The developed system is based on a very low-cost IoT SN equipped with multiple sensors. Some of these are simple with an analog application interface (e.g., temperature, Galvanic Skin Response—GSR, ECG, electromyogram-EMG sensors), while others are smart with digital application interfaces (like blood pressure, pulsioximeter (SPO2) and glucose sensors. An analog sensor is coupled to one of the multiple channels of the Analog Digital Converters (ADCs) embedded in the microcontroller board. Passive or active analog circuitry may be needed to adapt the output range of the sensor's values so as to match the ADC's input value range. On the other hand, Low Noise Amplifiers (LNAs) apart from Wheatstone bridges or voltage dividers could be employed to extend the output range. Regarding digital sensors, these are often calibrated to easily retrieve their data and send them to the IoT SN through a dedicated digital serial or parallel bus. They could also be coupled to the GW. Accuracy of the digital sensors generally cannot (and usually need not) be improved, but filtering could be applied for enhancing accuracy of the analog sensors and stability as elaborated in the rest of the paper.

The digital interface of the smart sensors can be connected to the Sensor Node via serial interfaces. However, conflicts could happen if e.g., SC functionalities running on the IoT SN and the GW may also require the same (serial interface) so as to be controlled. This results in conflicts concerning the sensors which have to be coupled directly to the IoT SN. General Purpose I/O (GPIO) could be utilized to connect sensors that exhibit simple communication interface but this may increase the operation overhead of the serial bus while additional SW drivers are required to be installed.

A conflict could also appear due to limited software resources available, like the bus drivers. If SW switching is permitted, some conflicts can be controlled by letting sensors operate in an alternate way, i.e., by selecting one software or hardware resource at a time. Extra HW switching could be required support for different sensors that are attached to the same I/O. Hardware switching can be implemented using relays, optocouplers, etc.; a solution which has also been adopted in our case. Auxiliary GPIOs are usually used to perform the HW switching, or in the case of a microcontroller's memory size not being large enough, RTOS (Real-Time Operating-System) scheduler (pre-emption and task prioritization) or semaphores could control access to the GPIO. The available HW resources of the microcontroller determines the storage capability of complicated firmware used for sensor data manipulation.

Libelium-Cooking Hacks provide a low-cost e-Health kit (v2.0) [13], which was chosen in the current work for the implementation of the IoT SN. This e-Health kit supports the following analog sensors: temperature, skin conductivity (Galvanic Skin Response—GSR), position detection, and breathing airflow sensor, as well as electrocardiogram (ECG) and electromyogram (EMG) ones (the last two sensors cannot be processed in the same time). The digital sensors supported include a sensor for blood pressure measurement that uses serial link communication, one for the measurement of heart pulses and the oxygen in blood (named pulsioximeter or SPO2 that requires 8 GPIO pins) and a glucometer. The glucometer and blood pressure sensors cannot be controlled in the same time since they share the same serial link (i.e., the Universal Asynchronous Receiver Transmitter—UART). An Arduino bridge is used to control the Arduino development board and the Raspberry PI one. Either one of them (Arduino or Raspberry PI) could alternatively form the required processing core for the operation of the e-Health kit platform. The scope of the Arduino bridge is to rearrange the pin-out of the Raspberry PI GPIOs and provide Arduino-compatible headers and support the shields that are already available for this platform. An additional ADC is also used to expand the number of analog inputs exhibited by the Raspberry PI. Although Libelium-Cooking Hacks has discontinued the version v2.0 of the e-Health kit, its successor (MySignals) is a much more expensive platform that would not be able to support a major target of this work: low cost. For this reason, we have developed techniques that enhance the sensors' precision and the support of lower power operation without increasing the cost of the overall system.

The Raspberry PI processor of the e-Health kit was chosen since its 32-bit ARM core offers higher processing power than the 8-bit AVR of Arduino. Furthermore, the increased Flash and RAM memory embedded in the Raspberry PI module allows for more advanced and complicated software to be developed. It also provides the ability for local processing of the (sensors') sampled data. The employed Libelium-Cooking Hacks e-Health kit v2.0 is not able to cooperate with other processors, although it would be desirable to employ a microcontroller that is computationally powerful but still offering low energy consumption options as AVR does. In a future version of our system, the low cost e-health kit circuits could be adapted to such a 32-bit low power microcontroller. A C-programming language library like the one currently offered by Libelium-Cooking Hacks for the support of Raspberry PI would have to be developed in such a case. Several of the techniques presented here such as sensor filtering, communication protocols between the SN and the Gateway, etc, would also be applicable to such a platform. Figure 2 delineates the specific way that the e-Health kit comprises both the Raspberry PI and Arduino bridge. This stacked connectivity eases the installation of wireless communication devices. This module could be a Wi-Fi, Bluetooth, GPRS, etc., and is used for enabling the communication link between the GW and IoT SN. The IEEE 802.11 b/g/n Wi-Fi protocol was adopted due to its connectivity reliability in comparison to Bluetooth devices. The top pin connectors of the e-Health as shown in Figure 2 have been employed to couple the aforementioned sensors.

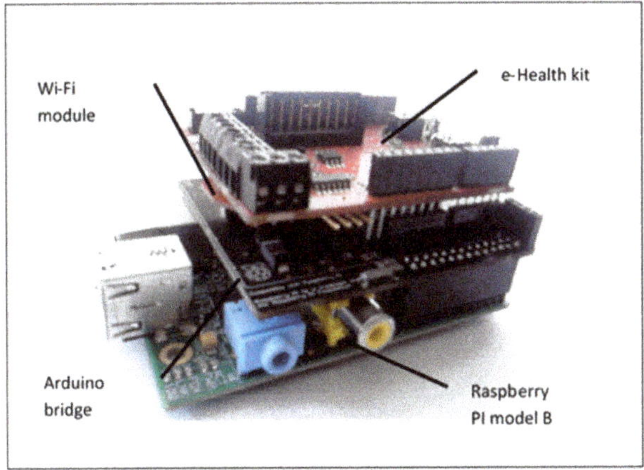

Figure 2. Stacking of the IoT SN modules.

Figure 3 shows the communication paths used between the proposed Internet of Things Sensor Node (IoT SN) and the Gateway when they operate in debug (Figure 3a) and real time mode (Figure 3b). As delineated in Figure 3, a secured FTP (sFTP) server (like FileZilla) has been installed on the Gateway side for the implementation of the Sensor Controller—Gateway communication protocol. This protocol is implemented through two types of files: (a) a command file which is prepared in the Gateway and downloaded by the Sensor Controller describing the sampling strategy and (b) the data file which is used to store the measured sensor values and is uploaded to the Gateway. In real time operation, the IoT SN is connected wirelessly to the Wi-Fi (IEEE 802.11g) router. The Gateway can be also connected in a wired manner to the router since it is not necessarily portable. In debug mode, the IoT SN is both wired and wirelessly connected to the router. The wired connection allows for the debugging of the

IoT SN by the Raspbian operating system (OS), or simply by a telnet terminal (like putty) that can be opened at the gateway or a different computer.

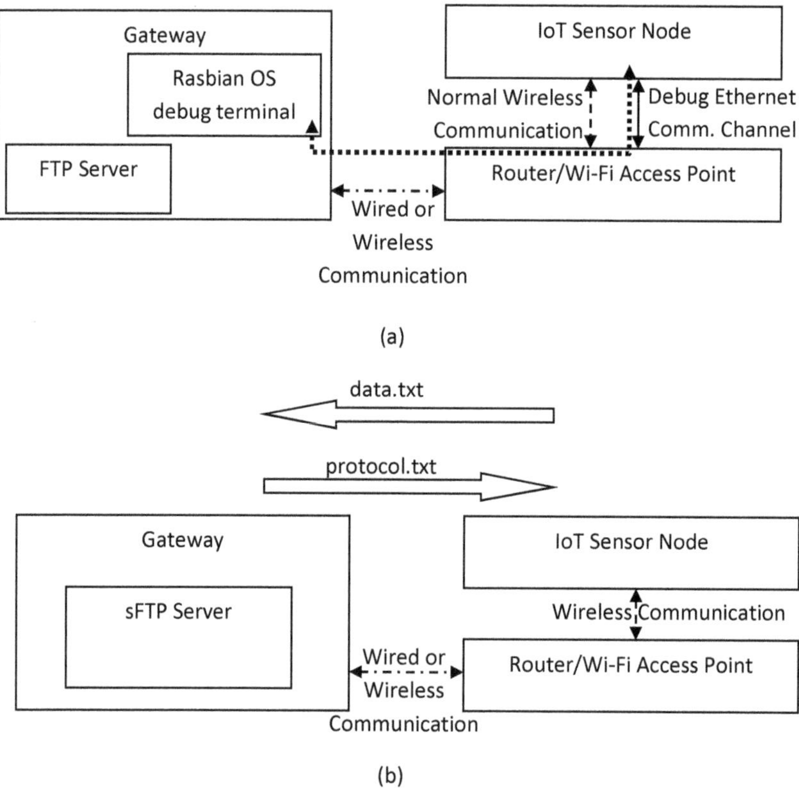

Figure 3. Sensor Controller and Gateway configuration in (**a**) debug and (**b**) real time mode.

Using the experimental set-up shown in Figure 2, we have focused on providing the specific platform with a number of value-added characteristics that could potentially make it competitive against much more expensive commercial e-health platforms. These key characteristics are analyzed in the following sections and include: significant enhancement of (analog mainly) sensors precision, increased portability by achieving low power consumption, and the development of a flexible and efficient communication protocol supporting all kinds of sensor sampling profiles and medical test scenarios.

2.2. Enhancement of Analog Sensor's Precision

As described before, the analog sensors are coupled to ADC inputs through customized range value adaptation circuits. Specifically, we have employed a 3V Wheatstone bridge to connect the temperature sensor to it, subject to on-board calibration. The single-ended signal is produced through the amplification of the differential signal of the output of the Wheatstone bridge. It is required to measure the real resistance values (R_a, R_b, R_c) of the Wheatstone bridge calibration scheme according to the manufacturer's suggestion. This has to be carried out with a multimeter that is also used to confirm the actual value of a reference voltage $V_r = 3V$. The actual values measured are used to overwrite the default ones set in the system software. Following this calibration scheme, any modification to the hard-written software parameters requires recompilation of the libraries' software. Moreover, the indications of various analog sensors are sensitive to environmental conditions like temperature, humidity, power supply, etc., and their outputs may be rippling.

In this work, we have tried to enhance the accuracy and stability of the measurements through real time software calibration and filtering. One way to achieve the software calibration is to modify and read in real time the parameters stored in configuration files. Sensor values can be exploited to adjust parameters that affect other sensors. For example, an increase in the measured temperature by the corresponding (temperature) sensor can be used to calibrate the GSR or the breathing airflow sensor. The calibration can be based on look-up tables that define target parameter values based in turn on other measurements. These tables can be either stored with the Raspberry PI Software in the SD-card, or can be downloaded from the sFTP server at boot time for easier update. External environmental conditions can be used as well, but although they may have been derived from high precision sensors, they may not reflect exact conditions in the room where the Sensor Controller resides.

The temperature sensor calibration referred to above is used as an example on how to modify the default calibration scheme proposed by the e-Health kit manufacturer. First, the parameters A_{ux} and R_{aux} are estimated as:

$$A_{ux} = \frac{V_t}{V_r} + \frac{R_b}{R_a + R_b},$$ (1)

$$R_{aux} = \frac{R_c \cdot Aux}{1 - Aux},$$ (2)

The estimated R_{aux} is exponentially related to the temperature as follows:

$$R_{aux}(T) = f \cdot p^T \Rightarrow T = \frac{log(R_{aux}(T)/f)}{log(p)},$$ (3)

The exact f and p values depend on the measured temperature. Equation (3) can be exploited for more accurate corporal temperature estimation, as well as for the calibration of other sensor indications that depend on temperature. The e-Health kit library checks whether R_{aux} lies within specific limits corresponding to draft temperature intervals of 5 °C in the range between 25 °C and 50 °C. The used values for f and p depend on the measured resistance R_{aux}. Even if the temperature sensor is not needed in some medical cases it can still remain connected to the IoT SN to roughly indicate the environmental temperature to be used for the real time calibration of other sensors.

The stabilization of the analog sensor indications can be performed by a filtering procedure that takes place at the Sensor Controller Software, which is executed on the IoT SN. Please note that the intention is not to develop a radically new filtering algorithm, but rather to present a number of well-known ones with different computational complexities in order to assess their efficiency on the developed system. Specifically, a Moving Window Average has been used with extreme value exclusion. The initial average value i.e., A_{init} is estimated by applying the moving average to the last k sample values $(v_t, v_{t-1}, \dots, v_{t-k+1})$:

$$A_{init}(t) = \frac{v_t + v_{t-1} + \dots + v_{t-k+1}}{k},$$ (4)

The k, v_i's are compared to A_{init} and the v_i's for which their difference from A_{init} is larger than the Th threshold, are dropped from the calculation of the final average computation. Alternatively, a fixed number of the k samples with the highest deviation from the initial average can be excluded. The associated binary index m_i is assigned to *0* for the dropped samples and to *1* for the remaining ones. The estimation of the final average A_{fin} is calculated by Equation (5).

$$A_{fin}(t) = \frac{m_t v_t + m_{t-1} v_{t-1} + \dots + m_{t-k+1} v_{t-k+1}}{m_{t-1} + \dots + m_{t-k+1}},$$ (5)

It may not be possible to apply the averaging method in a moving window of samples especially when other sensors have to be sampled in between. This is because a significant delay may intervene in some samples, thus making their sampling irregular (i.e., the samples are not retrieved in equal time intervals). In this case, k dedicated samples may have to be retrieved just to extract a single A_{fin} value. The time needed to retrieve these k samples may also not be negligible and has to be taken into consideration when the command file is prepared.

For instance, consider that one sensor must have to be successively sampled in 20 ms intervals and that $k = 5$. This means that an A_{fin} average can be extracted in 5×20 ms = 100 ms. If successive values have to be returned to the Gateway every 500 ms, then the sampling interval in the command file should be defined as 400 ms, since 100 ms more will be needed to estimate an average sample from this sensor. Of course, this procedure poses some restrictions on how small the sensors' sampling intervals can be.

More sophisticated methods like PCA [35] and Kalman Filters [36] have also been implemented and tested for higher sensor accuracy and avoidance of the delays caused by the Moving Window Average. PCA is used for extracting strong patterns in high dimensional data and can also be used to compress the data in a lossy way.

If $X_{rxn} = [X_1 \ X_2 \ \ldots \ X_n]$ is the measurements matrix, where X_i forms the sub-vector of r elements v_j, that are essential for the estimation of the principal components.

The covariance matrix of each X_i is calculated for the measurement of the data variability or data spread:

$$(R)_{rxr} = \sum_{i=1}^{r} X_i X_1^T, \tag{6}$$

The principal components transformation could be related to the singular value decomposition (SVD) of $(R)_{rxr}$:

$$(R)_{rxr} = USV^T, \tag{7}$$

where S is a diagonal matrix that contains the R singular values, V is a matrix that its columns correspond to the R right singular vectors while $U = [U_1 \ U_2 \ \ldots \ U_n]$ stands for the feature vector (matrix of vectors) with its columns being the R left singular vectors.

The reduced $(Ureduced)_{nxm} = [U_1 \ U_2 \ \ldots \ U_m]$, $m < n$ is created by the m sub-vectors from U i.e., the first m eigenvectors which are ordered by magnitude. If a low m value is selected, a higher compression rate is achieved (m/n), as well as a higher level of smoothing. The compressed data received are:

$$(Y_i)_{mx1} = (U_{reduced}^T)_{mxn}(X_i)_{nx1}, i = 1, 2, \ldots, n, \tag{8}$$

While the original data can be recovered by:

$$(X_i^{received})_{nx1} = (U_{reduced})_{nxm}(Y_i)_{mx1}, i = 1, 2, \ldots, n, \tag{9}$$

In the simplified Kalman filter used, the initial state x_0 is assigned to v_0. The confidence factor P_k (generally set to the squared standard deviation of the input values model) is initially set to $P_0 = 1$ (because the input values are not associated to a particular model). Similarly, we assign the measurement noise covariance R_n to be equal to 1. The iterative procedure of the Kalman filter updates the factors K_i and P_i as follows:

$$K = P_i/(P_i + R_n), \tag{10}$$

$$P_{i+1} = (1 - K)P_i, \tag{11}$$

$$x_{i+1} = x_i + K(s_{i+1} - x_i), \tag{12}$$

2.3. Sensor Controller's Communication Protocol

The fundamental communication scheme between the subsystems of the e-health system of Figure 1 is depicted on Figure 4. At some specific time, (t_0), the gateway reads the instructions concerning the case of the monitored person. If a set of medical tests is to be performed starting roughly at time t_2, the patient is informed by the tablet to get prepared (e.g., wear the sensors) a little earlier (at t_1). The instructions shown to the patient by the tablet are defined by the gateway. One of these instructions is to turn on the Sensor Controller, the software program executed on the IoT Sensor Node, and download the (sensors) sampling scenario by the gateway through the developed Sensor Controller-Gateway communication protocol. Acknowledgments (ACK) are exchanged to confirm that the patient has followed the instructions. The medical tests are performed in specific sampling intervals determined by the medical scenario. If these intervals are long enough, the Sensor Controller may enter a low power (sleep) mode to save power. When all of the tests defined in a scenario are completed, the sensor values are returned to the Gateway. Some local processing may be performed before the final values are uploaded to the Medical Database (DB). Meanwhile, the patient is informed that the tests have been successfully completed and he is instructed to turn off the Sensor Controller. Therefore, the communication mechanism of the protocol between the operating Sensor Controller and the Gateway is important for the support of a diversity of sampling strategies and thus, for supporting as many medical cases as possible.

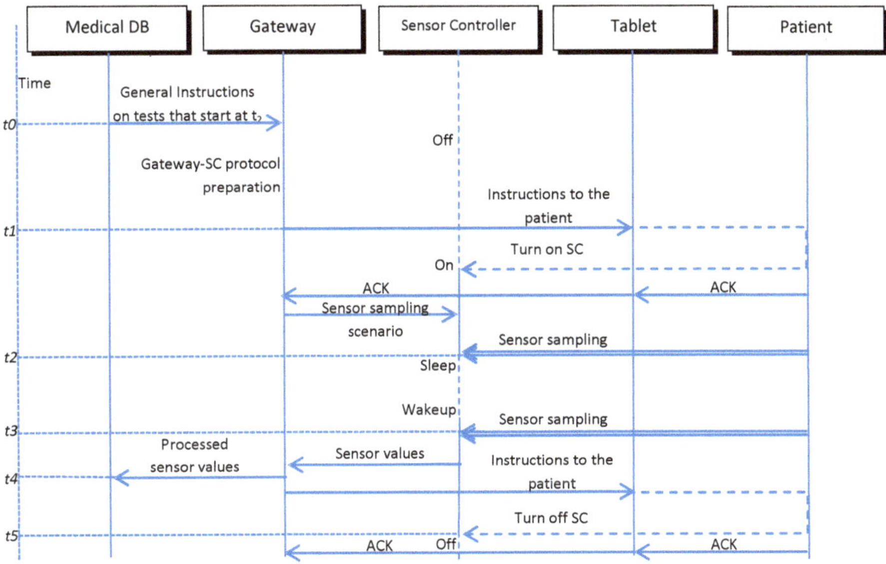

Figure 4. Typical communication scheme between the subsystems of the e- health monitoring system.

In addition to the communication scheme shown in Figure 4, the proposed IoT SN supports the following features: (a) initial filtering of sensor values within the Sensor Controller software to stabilize the measurements, (b) upload capability of the intermediate sensor values to the Medical Database of HIS, (c) variable sampling interval duration that ranges from milliseconds to hours in accordance with the various levels of power saving modes, (d) capability of performing several medical tests throughout a day or days. The communication between the GW and the Sensor Controller software is able to successfully cover any desired sensor sampling rate and medical test scenario. The communication protocol administrates the data and command interchange between these devices. Sensor Controller uses the commands to plan the potentially concurrent sampling of various sensors in different sampling intervals according to the specific medical scenario.

The privacy/security of the medical data is an important issue that has to be guaranteed at all levels in such an environment. Since we focus on the design of the IoT SN in this paper, we are interested in secure communications between the gateway and the software program executed on it (i.e., Sensor Controller). The privacy and security of the medical data stored on their way to the remote cloud as well as the privileges of the user that can access these data from the cloud (operators, medical staff, etc.) is beyond the scope of this work.

As mentioned in previous subsections, a secured FTP (sFTP) server is installed on the Gateway for the exchange of command and data text files replacing the FTP that had been employed in our initial versions where the only security was provided by the encryption offered by the Wi-Fi module. In sFTP, the transmitted datum is encrypted and a public key is sent to the remote terminal through Secure Shell (SSH). The authentication process includes a host key that is used to identify the sFTP server to the client and ensure that the server is known and trusted. The host key prevents man-in-the-middle security breaks and is part of the sFTP server configuration. After the sFTP server has been installed and configured on the Gateway side, the communication with the IoT SN is initiated. The employed message passing communication protocol is implemented by exchanging command and data files. Both of these files are transferred after the IoT SN initialization (the execution of the Sensor Controller software is started). By using the secure protocol, as shown in Figure 3, we ensure that data are transmitted via a secure path to mitigate possible cyber-attacks.

A command file defines either a sleep interval (in ms) or a specific sampling scenario. The IoT SN goes either into light or deep sleep mode if the sleep interval is less or higher than 30 s, respectively. If a sampling scenario is defined, two parameters are also specified: Q (Quantum) and I (Interval). The parameter I determines that after the sensor which exhibits the lower sampling rate has been read once, the Sensor Controller will re-read the command file after a delay of I ms. Then, the new command file that will be read might have been modified by the Gateway if a new sampling scenario has to be applied. The same sampling scenario will be performed by the SC, if the SW that is executed on the GW does not modify the command file.

In the command file, an integer number N is also defined that is associated with a sampling interval of $N \times Q$ ms, which is to be applied to a specific sensor. If N is equal to 0 the sensor is not used and is never sampled.

According to the specification previously defined, the SW that is executed on the SC downloads the command file and initiates a sampling loop of the sensors provided that N is non-zero. At the end of each one of the iterations, a delay of Q ms is inserted. A sensor is read if the current iteration number modulo its N value is zero. For example, let's assume that the parameter N is defined for the pulsioximeter, Airflow, ECG, temperature, body position, and GSR sensors as follows: $NBS = 5$, $NAF = 2$, $NEC = 1$, $NBT = 200$, $NPP = 10$, $NGSR = 10$, respectively. This means that the pulsioximeter sensor is read at the iterations numbered 5, 10, 15, etc. that correspond to 5×100 ms, 10×100 ms, 15×100 ms, etc., respectively if $Q = 100$ ms. Similarly, the ECG will be sampled at 100 ms, 200 ms, 300 ms, etc.; the airflow sensor will be sampled at 200 ms, 400 ms, 600 ms, etc. and so on. Both the Patient Position and the GSR will be read at 1 s, 2 s, 3 s, etc. The largest N value has been assigned to the Body Temperature sensor. This value may correspond to a dummy sensor read, since it will be read once and then after an interval of I ms a potentially different command file will be downloaded by the gateway.

It is obvious from the description above that the format of the command file can support any sampling scenario and interval of any duration. Furthermore, long sampling intervals could be exploited to trigger a transition to deep sleep mode for achieving lower power consumption. During shorter sampling intervals, the sleep commands of the operating system could be used that achieve slightly lower power consumption due to the reduced processor activity. Furthermore, the data samples are also stored in the text files and transmitted back to the sFTP server. These data files support the delivery of mixed sensor indications that are identified by the sensor name and the accompanied measured data value with timestamp. The timestamp that accompanies the data uploaded to the gateway can be exploited to detect missing data e.g., due to internet connection loss. In this case, the gateway can ask the Sensor Controller to upload again the missing measurements. Higher level protocols, or the supervising medical staff could order additional tests if some data seem to be missing, taking into consideration the timestamps. The GW can be programmed to separate the sensor values of different data files. Although the proposed communication protocol does not support the interoperability or extendibility of a digital sensor network like the one presented in [15], it can nevertheless handle analog sensors in a versatile way in the sense of supporting almost any sampling profile. Future work will incorporate a Message Queuing Telemetry Transport (MQTT) protocol to transmit data over a broker to the user so as to scale the architecture towards the support of multiple users [34].

2.4. IoT SN Control Circuitry and Power-Consumption Analysis

The power dissipation is another critical feature of the IoT SN stemming from the necessity for portability and the fact that it is powered by a light-weight battery that is required to last for a long time of period so as to minimize recharging rate. Therefore, energy-consumption of the IoT SN is a critical issue that should be as low as possible in a portable health monitoring platform, so that its usage proves to be efficient and easy to handle. As already mentioned in Section 2.1, the Raspberry PI platform is employed due to its higher computational power and memory that it offers in comparison with the other supported microcontroller, i.e., Arduino. However, Raspberry PI bears a problem regarding battery powered systems in the sense that it does not support sleep modes. For this purpose, special hardware extensions have been employed to overcome this drawback. The parameters that determine the power consumption of the IoT SN include: (a) the computational burden and the idle intervals of the microcontroller, (b) the power consumption of peripheral circuits like the wireless communication board, (c) the number of connected sensors, (d) other analog circuitry and glue logic existing on the Sensor Controller board like external ADCs, adaptation circuits (LNAs, bridges), etc. The control of the peripherals' power, the processing core computational burden, the processor sleep modes etc., constitute the high level features that can be exploited to reduce the consumed power. Furthermore, the microcontroller used in the IoT SN may exhibit some restrictions that inhibit optimal low energy operations, as discussed in other sections.

Within this context, the power consumption of the individual modules is analyzed as follows: The power consumption of Raspberry is associated with the specific versions used. Models B/B+ have been employed in the present configuration due to their reasonable power consumption (with respect to other versions) while in the same time they provide the additional Universal Serial Bus (USB) host ports required to communicate with the Arduino and test various USB Wi-Fi modules. The PureLink/High Definition Multimedia Interface (PL/HDMI) interfaces are able to be switched off to save approximately 20 mA since it is not used at all in the present application.

The Wi-Fi module (RN-171 model) exhibits a sleep mode with power consumption of only 4 μA. This module enters into sleep mode automatically. One counter sets the sleep mode time interval set, while another wakes the device after another time interval. The maximum time delay for the Wi-Fi to enter/recover from sleep mode is estimated to be several hundred hours. The device can enter the sleep state either being in UDP or TCP modes. The receive (RX) and transmit (TX) paths of the Wi-Fi module consume 35 mA and 185–210 mA, respectively, for transmission power ranging from 12–18 dBm. The vendor of the e-Health platform does not include any reference to the specific power consumption of the kit or the Arduino Bridge; it is merely mentioned that they are supplied with 12 V/2 A. However, power-consumption tests have shown that in fully operating mode the measured current is 1 A. Therefore, if an external 1000 mAh battery is used and the IoT SN is continuously active, it will requires recharging every hour.

A large power amount is also dissipated by the (Arduino, Raspberry PI) boards' voltage regulators of the e-Health kit. The Raspberry PI regulator is bypassed while the input of the Arduino regulator is set to 6 V, which is marginally the minimum input voltage that can be used. The small variance between the 6 V input and 5 V output of this regulator reduces thermal power losses.

The basic improvement that allows for the exploitation of idle intervals between measurements is to use Arduino to control the processor's (Raspberry PI) sleep mode since this Raspberry PI processor core is not equipped with sleep mode capability. This is feasible due to the fact that measurements are taken after the elapse of long time intervals. Therefore, the IoT SN could enter the sleep mode switched-off during these idle intervals, and activated during the next measurement timing point. The implementation of such a configuration is presented in Figure 5, where the Arduino board is controlled by Raspberry PI through USB. Arduino controls, in turn, the power supply of the Raspberry PI as well as the (ECG/EMG) sensor software switching through a low power relay configuration. The power control of Raspberry PI is performed by Arduino according to the instructions received by the Raspberry PI through the USB port. According to Figure 5, Arduino is constantly connected to an external battery and a relay that is turned on (which is normally off) is used to power-up Raspberry PI. If Arduino is turned on, its SW sets on the relay to power up Raspberry PI. Additionally, Arduino is able to switch off Raspberry PI for a specified time period. In this case the Raspberry PI program is responsible for informing Arduino to perform the aforementioned action. Arduino's microcontroller consumes less than 1 mA in normal operating mode and approximately 0.1 μA@1.8 V in power down mode. Two relays are connected to Arduino's digital pins, which dissipate a current of approximately 50 mA each. Consequently, it is deduced that the total power consumption of Arduino's board, along with the relays, is approximately 101 mA during normal operating mode. On the other hand, Arduino's energy consumption is really insignificant while in power-down mode.

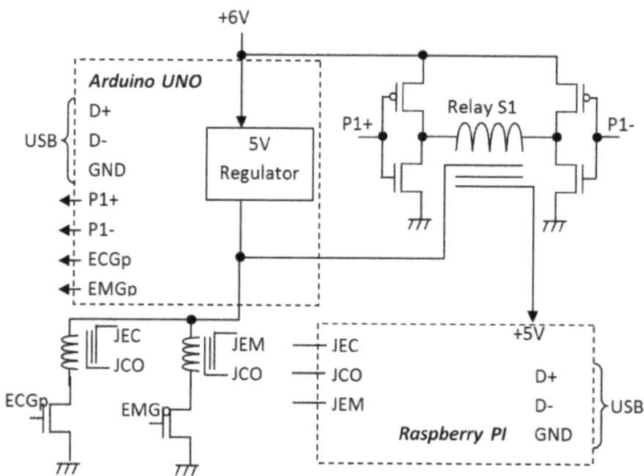

Figure 5. Block Diagram of Arduino and external relays for sensor selection and sleep intervals implementation.

Regarding the relay's power consumption, it should be mentioned that in order to avoid a constant current flow through the relay S1 (see Figure 5) either when it remains open or closed, the S1 was selected as a pulsed on-off type of relay. These relays change state when current flows for a short time interval to a specific direction. Then, when the current flow stops, it remains in this state until there is a current flow of opposite direction in the relay coil. The specific relay used requires a current of 100 mA for about 150 ms to change state. The current in this relay is controlled by four MOSFETs connected as a pair of CMOS inverters. Two digital pins (P1+, P1−) are used from Arduino for the control of this relay. The current drawn from these pins is negligible since the gate leakage of the MOSFETs is in the order of n A.

The e-Health kit has a 3-pin jumper (*JEC*, *JCO*, *JEM*) for connecting to either the ECG or the EMG sensor, since these sensors cannot be both used at the same time. In general, this is not a problem since they are targeted to different types of applications: ECG is a cardiac diagnostic test while EMG is used to control prosthetic limbs. When the common pin *JCO* of the 3-pin jumper is connected to *JEC*, the ECG sensor can be used, otherwise the *JCO* must be connected to *JEM* for enabling the use of the EMG sensor. Very low power relays can be used to control the connections of this 3-pin jumper in real time by software. Two digital Arduino pins are used to control these relays (*ECGp* and *EMGp*). The power consumed can be further reduced if optocouplers are used instead of magnetic relays or if a circuit like the one used for the control of the relay S1 is adopted.

When Raspberry PI is initially powered, a boot script is executed that starts the software of the health monitoring system. One of the first tasks performed by this software is the downloading of the command file from the file server installed at the gateway. This command file defines the sensor sampling profile. Intermediate sampling intervals can be defined that range between milliseconds and hours, as will be described in the next section. When the sampling interval is in the order of a few seconds, the only way to save power is to deactivate Raspberry PI peripherals that are not used, such as the wireless interface (light sleep). However, if the sampling interval is longer than 30 s, it is worth turning off the Raspberry PI completely (deep sleep) during consecutive samples. This is accomplished by sending an appropriate instruction to Arduino before Raspberry PI shuts down. Arduino waits for a while to let Raspberry PI completely shut down and then safely turns off its power through the relay S1 of Figure 5. Although each one of the Arduino's digital pins can pull up to 50 mA, the power consumed on these pins is not taken into consideration, since they are connected to MOSFET gates, as already mentioned.

The ATmega1280 microcontroller of Arduino can remain in the supported sleep mode for a maximum of about 9 s, where it consumes approximately 100 µA. Long intervals of deep sleep mode can be supported as multiples of these 9 s intervals using the EEPROM of the ATmega1280 for storing a counter. At power up, this counter is cleared. When Arduino awakens from a 9 s sleep interval, it updates this counter and falls asleep again until the desired overall sleep interval is completed. In this case, the Raspberry PI is turned on again. When Raspberry PI is turned on, the health monitoring firmware recognizes that a deep sleep interval was completed (through the configuration information that was stored before it shuts down). In this way, it can continue sampling as instructed by the Gateway until a new sampling strategy is defined. The configuration information stored in the SD-card of the Raspberry PI and the EEPROM of the Arduino can be exploited to recover from power loss. In real time the Raspberry PI can also send commands to Arduino through USB that select either the ECG or the EMG sensor. Thus, the USB communication protocol between the Raspberry PI and the Arduino supports three commands: (a) the Sleep command where the Raspberry PI gets in sleep mode (b) the ECG select command and (c) the EMG select command. The Arduino replies with an appropriate acknowledgement.

Figure 6 shows the power consumption in various real time operating modes. It has been assumed that the total power consumption of the Arduino Bridge, the Raspberry PI, and the e-Health module is 700 mA after bypassing the voltage regulators. The PL/HDMI interface is switched off. The average power consumption of the Wi-Fi module is assumed to be 200 mA and is not included in the 700 mA consumption mentioned above. The Wi-Fi module is turned off during short idle periods (light sleep) but its "on" operation could be additionally limited to short time intervals, taking into account the time-schedule where data is sent to and received from the GW. Arduino's microcontroller draws approximately 1 mA in normal mode and less than 1 µA in sleep mode. When Arduino awakens to perform updates to the EEPROM sleep counter, 1 mA is consumed by the microcontroller, and this operation can be completed in less than 10 ms (required for one read and one write to the EEPROM). The relay S1 of Figure 5 draws 100 mA for 150 ms when Raspberry PI is turned on or off. Consumption of the relays controlling the ECG/EGM sensors will not be taken into consideration since most of the time they will be in their default state: Normally Open (*NO*) or Normally Close (*NC*) state where no current is drawn. Even in the rare case where the non-default sensor (ECG or EMG) will be selected, the power consumption of the relays is less than 50 mA or completely negligible if a connection like that of the relay S1 is used.

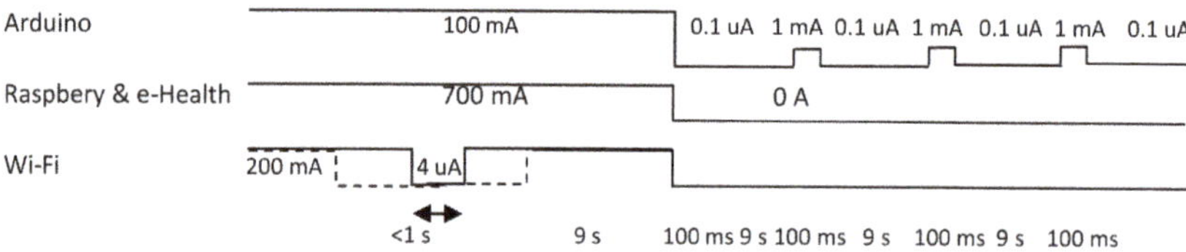

Figure 6. Power Consumption during normal operation and sleep modes.

At this point we would like to note that, as it is easily implied, our sensor node from time to time will have to be connected to an AC power adapter for recharging purposes. For this reason, all necessary safety measures have been taken to avoid possible accidents (of the monitored persons), such as the employment of suitable safety relays, special fusions, etc.

2.5. Case Study

An extended schedule of medical tests is examined to demonstrate the capabilities of the proposed IoT SN. The energy consumption improvement is compared to the energy that the system would dissipate if the presented power enhancement techniques had not been employed. The power savings always depend on the specific medical scenario that is examined. This is the best comparison we can provide since it is not easy to compare the power of the developed system with others that operate under similar conditions. The framework where this medical scenario can take place is described in Figure 1. Specifically, each one of the IoT SNs are placed next to the medical beds in the medical care premises (e.g., rehabilitation center, hospital), the patient wears the sensors that are attached to the IoT SN while physicians can either locally or remotely access the data provided by the sensors (either through the local Gateway or via cloud). The actions that take place at each step are compatible with the communication protocol described in Figure 4 and are described extensively in this section. The estimation of the reduced power consumption by employing the defined sleep modes is also given. Within this context, an indicative medical scenario to be demonstrated is the following: Initially, the person position is tested four times every minute to verify that the proper position has been taken for the following tests: an ECG test is executed by reading the ECG sensor for 2 min and 100 ms intervals (no averaging filtering was used), then a 10 min break gives the person or the nurse the opportunity to disconnect the ECG probes. After that, the airflow and the galvanic skin conductance sensors are sampled for 2 min (with filtered averaging samples $k = 5$ and 20 ms sampling interval). Samples of the airflow and the galvanic sensors should have 200 ms and 500 ms time distance, respectively.

The following actions have to take place in order to implement this medical scenario: A command file with a sleep instruction is sent to the sFTP server working directory from the last measurement while Raspberry PI (if not completely turned off) awakens, retrieves this file, and is switched to the sleeping mode again until a different than sleep command appears in this file. The patient is instructed by the tablet to wear the patient position sensor and the ECG electrodes.

The patient wears the body position sensor and the ECG electrodes, turns on IoT SN, and informs the gateway through the tablet or smart phone for his availability to start the test. The gateway changes the command file defining $Q = 1000$, $I = 0$, $NPP = 60$, $NBT = 240$. When this file is downloaded from the GW, the patient position sensor starts to be measured immediately by the Sensor Controller software every minute for $NBT \times Q = 240 \times 1000$ ms $= 240000$ ms $= 4$ min until a dummy read is performed to the temperature sensor. Then, the measured data concerning the patient position are uploaded and the next command file is downloaded from the FTP server.

In the previous 4 min interval, a new command file is prepared at the GW for the 2 min ECG test: $Q = 100$, $I = 6000$, $NEC = 1$, $NBT = 1200$. If this is not changed within the initial 4 min interval, the body position of the patient will be monitored for 4 more minutes before the ECG test starts. After the ECG's test completion, a dummy read is performed of the body temperature sensor, and the ECG data is sent to the GW. However, the new command file is going to be read after 10 min, as defined by the last I and Q parameters. During this 10 min sleep interval, the patient is instructed by the tablet to remove the ECG electrodes and wear the airflow and galvanic skin conductance sensors.

In the next command file issued by the gateway within the 10 min sleep interval, the following parameters are defined: $Q = 100$, $I = 6000$, $NAF = 1$, $NGA = 2$, $NBT = 480$. It is noticed that the sampling intervals and the dummy read have been adapted appropriately, taking into consideration that 100 ms are required for the averaging of the sensor values. More specifically, the airflow sensor is sampled in every iteration but requires an extra 100 ms for performing averaging so as to estimate a single measurement. Taking into consideration that Q is equal to 100 ms, the airflow sensor generates one sample every 200 ms. Similarly, the GSR is sampled every two iterations, i.e., every 400 ms, but a delay of 100 ms is also inserted to take into account the time needed for its own averaging. This delay makes the sampling intervals of the airflow sensor irregular, in the sense that this sensor will be sampled, e.g., at 100 ms, 300 ms, 600 ms, 800 ms, etc. Regular sampling time intervals can be achieved if both sensors are sampled concurrently at the same rate, although some galvanic sensor values will be redundant in this case. The dummy read is carried out after 480 iterations. If r iterations are required for an interval of 2 min, then

$$r(100 \text{ ms} + 100 \text{ ms}) + \frac{r}{2}100 \text{ ms} = 120{,}000 \text{ ms} \Rightarrow r = \frac{2400}{5} = 480, \tag{13}$$

The first term in (13) takes into account the 100 ms delay of the iteration and the 100 ms delay of the airflow averaging while the second term is the delay of 100 ms for the galvanic sensor's averaging. Although these estimations may seem complicated, most cases do not need such precision for when to stop the sampling. The irregular sampling intervals could also be corrected at a higher level, e.g., through interpolation. The last I parameter defined (6000) calls for a 10 min sleep after the completion of the two sensors sampling. In this interval, the IoT SN can be turned off by the user if instructed to do so by the tablet. Otherwise, command files with sleep commands should be used to keep the IoT SN in low power operation.

In Figure 7, the various sampling and sleep intervals are shown as defined in the case study described earlier. Taking into consideration the power consumption during the normal sampling operation, i.e., the light, deep sleep, and shutdown modes, draft estimation can be performed on the energy consumed during this scenario. Specifically, in the first four minutes when the patient position sensor is sampled 4 times and each time the system goes to 1 min deep sleep mode, we can assume that for 4×30 s the power consumed is 1 A (shutdown mode). The power consumed during deep sleep mode is negligible and won't be considered. The system also enters deep sleep mode after the 2 min interval where ECG is sampled, and after the 2 min interval where the breathing airflow and the galvanic skin conductance sensors are sampled. Thus, the corresponding 1 A power consumption has to also be taken into consideration for an additional 2×30 s.

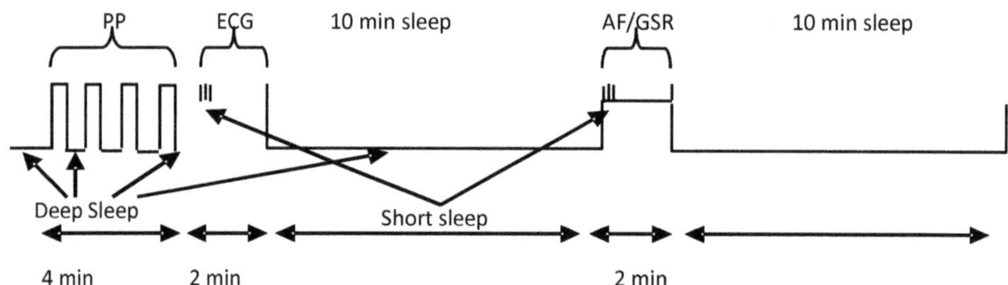

Figure 7. Description of sampling and sleep modes during the case study.

During the 2 min of ECG sampling, the power can be assumed as equal to 700 mA, which is raised to 900 mA only when the communication with the gateway takes place (the Wi-Fi module can be turned on only during the data/command exchange). Regarding power estimation, we will assume that the Wi-Fi is active for an average duration of 10 s and that the overall Sensor Controller consumes 900 mA during this interval.

Similarly, in the two minute interval used for the airflow and the galvanic skin conductance sensors' sampling, the power consumption can be assumed to be 700 mA since the averaging does not affect the consumption if the Wi-Fi is inactive (light sleep mode). Then, it is raised to 900 mA for about 10 s when the communication with the Gateway takes place for uploading the sensor data. Consequently, the overall power consumption is:

P1 = 4 × 30 × 1000 + 120 × 700 + 10 × 900 + 30 × 1000 + 120 × 700 + 10 × 900 + 30 × 1000 = 120,000 + 84,000 + 9000 + 30,000 + 84,000 + 9000 + 30,000 = 282,000 mA·s = 78.3 mAh

If a full power operation of 900 mA had been used during the 18 min interval of the medical scenario described above, the consumed power would have been P_2 = 972000 mA·s = 270 mAh. Thus, an overall saving of 70% is achieved. If the afore-described scenario had to be performed every day, then even if the Sensor Controller was turned off before and after the tests, a 1000 mAh battery would require recharging in less than 4 days while the proposed scheme can extend the recharging interval to more than 12 days. Of course, power saving depends on the specific medical scenario as already mentioned. The energy consumption can vary considerably among various medical test-cases and sampling profiles, but the previously mentioned test scenario could be considered as indicative of the power savings that could be achieved when the proposed techniques are applied.

3. Results

This section presents the experimental results regarding the various sensors of the proposed low-cost IoT SN, as well as their comparison, where possible, with expensive certified sensors. The tested sensors mainly include the Galvanic Skin Response Sensor (GSR), the Body Temperature Sensor-(BT), the Airflow Sensor (AF), the Patient Position Sensor (PP), and the sensor for Electromyogram (EMG).

It should be mentioned that the e-Health kit of Libelium which has been used as the basis for the development of the IoT SN presented here, also supports the digital sensors for Blood Pressure (BP) and glucose measurement. However, they could not be used because these sensors, together with the Wi-Fi communication board, are required to be connected to the same physical resource: the serial port. Given that the communication of the IoT SN with the gateway is performed via Wi-Fi, it is obvious that these sensors could not be included in this evaluation procedure

Moreover, the digital pulsioximeter (SPO2) sensor is autonomous with its own display. The IoT SN is simply used to transfer the measured results from the particular sensor to the gateway without any change in accuracy of the measured values, and so it had no meaning to test it further. Regarding the remaining sensors that were tested, we attempted to provide an objective evaluation concerning their measurements. The results of some of them could be easily checked and evaluated, such as the Patient Position Sensor one. However, this is not always possible. For example, the airflow and the EMG sensors cannot be evaluated by comparing absolute values. It is sufficient though, to check whether these sensors can follow the breathing or muscle contraction patterns. Therefore, in this case, it is not required to compare their output with the output of similar commercial devices. However, such a comparison is performed whenever possible, like in the case of the temperature sensor, which was evaluated against a commercial medically certified one. Regarding the GSR sensor, an attempt was made to match the peaks of its output with another medically certified commercial one by Shimmer [19]. In the following, the evaluation-testing results of the various sensors are given, along with a number of conclusions that were derived from this procedure. It has to be stressed that the presented sensor curves are representative of numerous identical experiments carried out with similar results. This holds for all the different types of experiments described.

In Figure 8, the temperature sensor values are evaluated against the reference ones as provided by a certified thermometer and the values that are produced after the application of the filtering algorithms described in Section 2.2. The Moving Average Window is applied for $k = 5$ samples that are initially averaged, and two of these samples are excluded (the ones with the highest deviation from the initial average). The PCA algorithm was applied using $m = 3$ columns of the $n = 10$ of the U matrix (compression 30%). The Kalman filter is applied following the exact algorithm described in Section 2.2. The same filtering algorithms and configurations have also been used for the GSR sensor in Figure 9. The Mean Square Error (MSE), the Normalized MSE (NMSE) and the Signal to Noise Ratio (SNR) parameters were mainly used for comparison. The aforementioned parameters are defined for a reference signal v_{ref} of N samples and a real signal v according to Equations (14) and (15).

$$MSE\left(v, v_{ref}\right) = \frac{1}{N} \sum_{i=1}^{N} \left(v(i) - v_{ref}(i)\right)^2, \tag{14}$$

$$NMSE\left(v, v_{ref}\right) = \frac{MSE\left(v, v_{ref}\right)}{MSE(v, 0)}, \tag{15}$$

$$SNR = 10log_{10}(\frac{v_{ref}}{v - v_{ref}})^2, \tag{16}$$

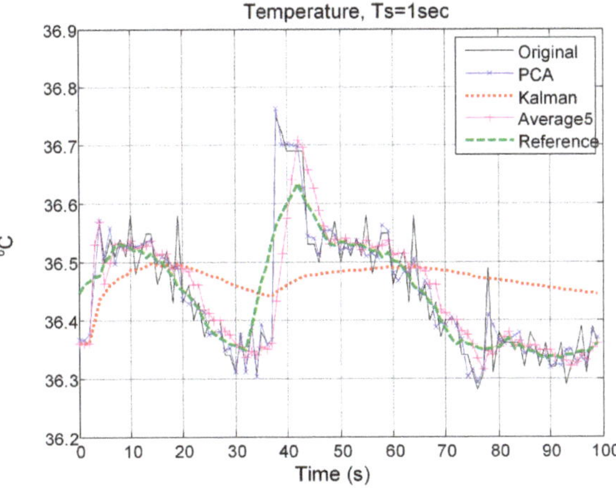

Figure 8. Temperature measurement with sampling period 1 s.

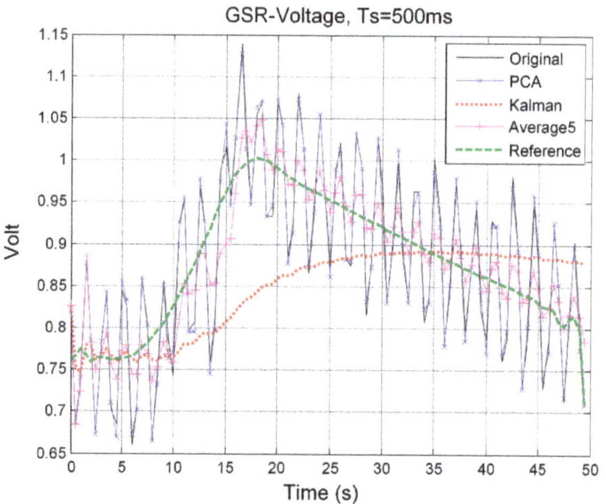

Figure 9. Skin Conductance Voltage Measurement with sampling period 500 ms.

Table 1 depicts the MSE, NMSE, and SNR parameters for the signals illustrated in Figures 8 and 9. Even though the skin resistance (R_s) and skin conductance ($1/R_s$) are used rather than the skin conductance voltage (V_s), these values are simply estimated taking into account the skin conductance voltage according to the formula $R_s = 50\ \text{k}\Omega/\text{V} - 0.5$ V, as in our case.

Table 1. MSE/NMSE/SNR parameters' comparison for Figures 8 and 9.

	Results	MSE	NMSE	SNR (dB)
Temperature (Figure 8)	Original	0.0032	2.37×10^{-6}	129
	Averaged	0.0019	1.42×10^{-6}	134
	PCA	0.0023	1.71×10^{-6}	132
	Kalman	0.0075	5.65×10^{-6}	120
Skin Conductance Voltage (Figure 9)	Original	0.0048	0.0062	51
	Averaged	1.42×10^{-6}	0.0012	67
	PCA	0.0049	0.0063	51
	Kalman	0.0064	0.0082	48

It is deduced from Figures 8 and 9 and Table 1 that the usage of the Moving Window Average filtering algorithm on the original (sensor) values has improved them significantly when compared to the reference measurements. Regarding application of the PCA algorithm, in most cases reduces the error and compresses the signal. Moreover, if a higher compression rate had been selected (with a lower m value) an even lower error would have been achieved in the particular examples where the smoothing is preferable than the preservation of the original signal details. At this point we would like to mention that, to our knowledge, this is one of the few works where PCA is used mainly for noise reduction of biometric signals. Undoubtedly, there exists a plethora of works in the literature where PCA techniques are applied to various biometric sensors (ECG, EEG, EMG etc.), like in [37,38], but their exclusive target in most of these cases was to achieve compression and classification. In the last few years the efficiency of PCA in noise reduction was acknowledged and it has been proposed to be used in applications like noise reduction for Doppler radars [39] and in SNR improvement for Inchoate Faulty Signals [40].

The application of the Kalman filter was not successful since a higher error appears. This error could have been reduced by tuning appropriately the parameters P_0 and R_n defined in Section 2.2. However, a different tuning would have been required for different sensors. In Figure 10, a pair of

GSR sensors was connected to different users responding to the same stimuli. As shown from this figure, the resistance pattern was quite similar in the two plots in terms of local peaks (minima and maxima), taking into consideration that the User 1 responds about 1 s earlier to the same stimuli than User 2. The dashed lines in Figure 10 show the peak correspondence between User 1 and User 2. However, the absolute resistance values corresponding to the two users and their amplitude are significantly different.

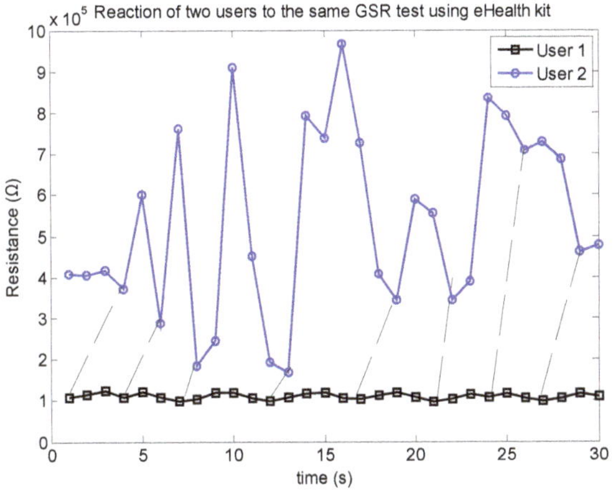

Figure 10. Comparison of two identical Libelium GSR sensors attached to different users that are subject to the same test. The vertical dashed lines denote the corresponding local minima between the two users.

Figure 11, shows the resistance variation of a Libelium e-Health GSR sensor and a Shimmer one, attached to the same user that is subject to a Stroop test. The difference in the absolute values of the two plots should not be taken into consideration. The important issue noticed here is that the patterns are similar and most of their peaks match as shown by the dashed lines. This is an indication that the GSR sensor of the developed e-health system operates reliably.

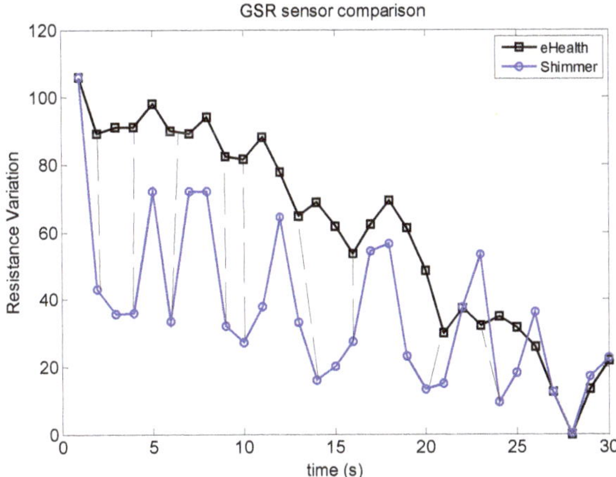

Figure 11. Comparison of the Libelium and the Shimmer GSR sensors behavior on the same person. The vertical dashed lines denote the corresponding local minima between the two sensors.

In Figure 12, the airflow sensor of the developed system is tested with a pattern of two deep breaths followed by two moderate ones. The sensor was used without applying filtering (Section 2.2) and as a result of this, additional peaks appear. In Figures 13 and 14, the breathing pattern used is alternately a deep and a moderate breath. A slightly different type of averaging is used in these figures. Filter *type 1* depicted on Figure 13 extracts the average by employing three successive samples. Filter *type 2* shown in Figure 14 follows the averaging method described in Section 2.2 with 5 initial samples, and rejecting 2 extremes for estimating the final average. As it is derived by these figures, both averaging schemes represent clearly the breathing pattern used. The rippling at the end of the curves is due to the fact that the sensor is removed at that time instant from the user.

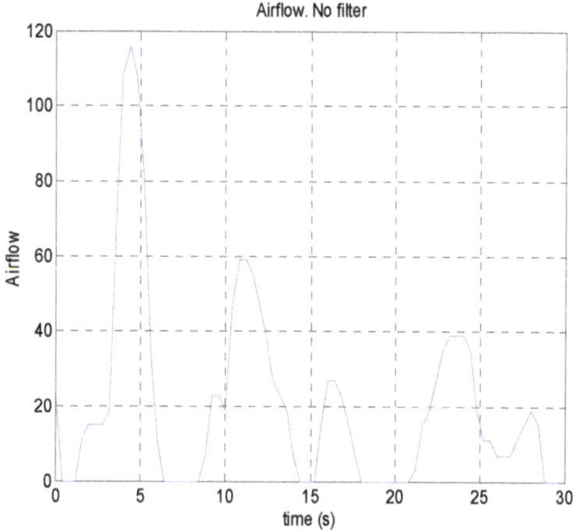

Figure 12. Airflow sensor tested; the user takes two deep breaths followed by two moderate ones. No filtering is used.

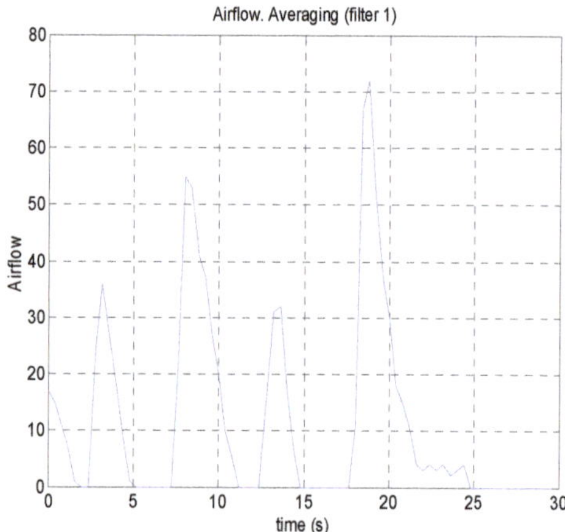

Figure 13. Airflow sensor tested; the user takes alternately a deep and a moderate breath. Filter *type 1* was applied.

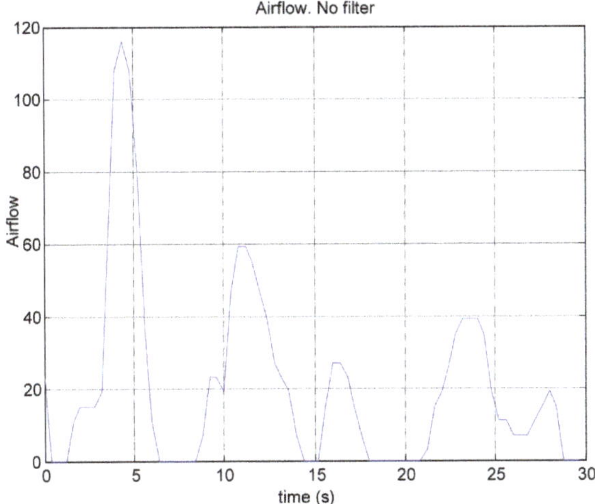

Figure 14. Airflow sensor tested; the user alternately takes a deep and a moderate breath. Filter *type 2* was applied.

The EMG sensor's results are shown in Figures 15 and 16. In Figure 15 a relatively small sampling period of 200 ms was used with no filtering. A moderate muscle contraction is followed by a strong and then a small one. Although an intense rippling appears, the muscle contraction pattern is recognizable. The same muscle contraction pattern is used in Figure 16 with a longer sampling interval of 1 s that allows the various filtering types to be applied. The reduced rippling allows the three muscle states to be recognized more clearly when averaging is used and especially when Filter *type 2*, of Section 2.2, is applied.

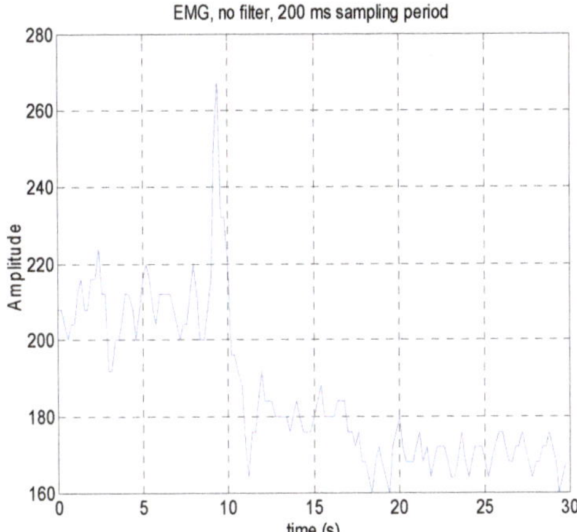

Figure 15. EMG measurement: a moderate muscle contraction is followed by a strong and a loose one. 200 ms sampling period is used with no filtering.

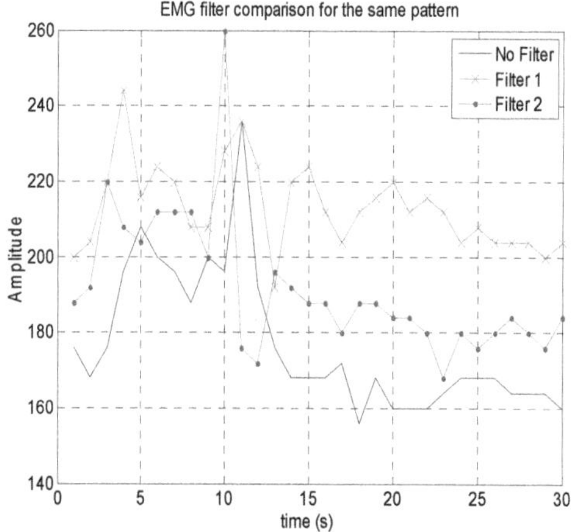

Figure 16. EMG measurement with the same muscle contraction pattern as in Figure 15. A sampling period of 1 s is used with various filtering options.

4. Discussion

Based on the aforementioned measurement results of the low-cost sensor controller platform, the following lessons were learned and conclusions drawn:

i. The sensors that can operate reliably are those that do not require a high sampling rate i.e., the temperature sensor, the airflow sensor, the patient positioning sensor, and the Electromyogram one. Simple filtering techniques can be applied with success, such as simple averages of the measured values (Filter *type 1*) or averages derived after discarding excessive values (Filter *type 2*).

ii. The Galvanic Skin Response Sensor, when compared with the corresponding medically certified sensor (from Shimmer), does not exhibit the same voltage levels, something which was at a certain degree expected. Nevertheless, the important issue is that the local peaks of their waveforms (which carry the essential measurements information) coincide at a percentage of 85%.

iii. The application of the averaging methods along with PCA and Kalman filter to a temperature and a GSR sensor showed that the error can be improved up to 6 times.

iv. The use of the PCA filtering algorithm can improve the error and compress the sensor data

v. The breathing airflow sensor seems to closely track the patient's breathing represented by the shape and width of peaks in the graph formed by the measured results.

vi. The patient position sensor is absolutely reliable, detecting with 100% success the actual (patient) position (no specific testing diagrams were included for this sensor due to their trivial nature)

vii. The Electromyography diagram produced by the corresponding sensor represents the intensity of muscle contractions in a fairly clear way. It is more preferable to focus on the differences of the successive measurement values instead of the absolute ones in this graph.

Furthermore, in order to avoid confusion, it is useful to apply filtering methods with a relatively slow sampling rate so as not to result in rapid fluctuations that are not related with muscle contractions. Future work includes the deployment of multiple IoT SNs and incorporation of the MQTT mechanism between Gateway and those Nodes. An IoT platform with database could be also used within the Gateway to perform further analysis of the received sensor values and handle all critical alerts on premises.

5. Conclusions

A low-cost and low power IoT Sensor Node with advanced measurement precision for e-health application that enables continuous patient monitoring has been presented. The sensor node is based on the e-health kit from Libelium-Cooking Hacks, which can perform various medical checks ranging from temperature and glucose level measurement to ECG tests. We have enhanced the key characteristics of the operations of the Raspberry PI single-board computer (model B/B+), which constitutes the basis for the aforementioned e-health kit, by improving the accuracy, through the incorporation of filtering algorithms to sensors' values within the implemented Sensor Controller. A flexible communication mechanism, which enables the definition of any type of sampling of sensors at the desired frequency, is also employed for successful data transmission and Sensor Node control. Furthermore, a power consumption minimization scheme has been developed by imposing suitable sleep intervals to the IoT Sensor Node (corresponding to the associated idle intervals between sensors' successive measurements) and periodically deactivating specific functions of the node (thus, expanding its battery life-time and autonomy from hours to days). A representative medical test-case was demonstrated and explored, elucidating the efficiency of the presented key features of the e-health IoT Sensor Node. These features, in combination with the successful testing/evaluation of most of its sensor functions, paves the way for further evolvement of the developed system to a low-cost e-health monitoring platform with capabilities comparable to high cost commercial counterparts. In this regard, future improvements encompass the utilization of the MQTT protocol that employs shorter TCP header and thus in turn will lead to even more lower power consumption and communication overhead for the IoT Sensor Node.

Author Contributions: Conceptualization, N.P., F.G. and M.B.; methodology, N.P., M.B. and F.G.; Software, N.P.; validation, N.P., M.B. and F.G.; project administration, M.B., writing—original draft preparation, N.P.; writing—review and editing, N.P., F.G. and M.B.; investigation, N.P., F.G. and M.B.; resources, F.G.

Funding: This research was partially funded by ELTAB project, grant number 46543.

Conflicts of Interest: The authors declare no conflict of interest.

References

1. Sadri, F. Ambient Intelligence: A Survey. *ACM Comp. Surv.* **2011**, *43*, 1–66. [CrossRef]
2. Dishongh, T.J.; McGrath, M.; Kuris, B. *Wireless Sensor Networks for Healthcare Applications*, 1st ed.; Artech House Publishing: Norwood, MA, USA, 2009; ISBN 1596933054.
3. Zhang, Y.; Sun, L.; Song, H.; Cao, H. Ubiquitous WSN for Healthcare: Recent Advances and Future Prospects. *IEEE Internet Things J.* **2014**, *1*, 311–318. [CrossRef]
4. Dang, W.; Manjakkal, L.; Navaraj, W.T.; Lorenzelli, L.; Vinciguerra, V.; Dahiya, R. Stretchable wireless system for Softwareeat pH monitoring. *Elsevier BV Biosens. Bioelectron.* **2018**, *107*, 192–202. [CrossRef] [PubMed]
5. Islam, S.M.R.; Kwak, D.; Kabir, M.H.; Hossain, M.; Kwak, K. The Internet of Things for Health Care: A Comprehensive Survey. *IEEE Access* **2015**, *3*, 678–708. [CrossRef]
6. Gay, V.; Leijdekkers, P. Health Monitoring System Using Smart Phones and Wearable Sensors. *Int. J. ARM* **2007**, *8*, 29–35.
7. Blood Pressure Monitors and Telemedicine Products. Available online: http://www.andmedical.com.au/ (accessed on 22 December 2018).
8. AliveCor KardiaMobile ECG for Phone and Android, Alive Technologies. Available online: https://www.alivetec.com/pages/alivecor-heart-monitor (accessed on 20 December 2018).
9. Chan, V.; Ray, P.; Parameswaran, N. Mobile e-Health monitoring: An agent-based approach. *IET Commun. Telemed. E-Health Commun. Syst.* **2008**, *2*, 223–230. [CrossRef]
10. Mukherjee, S.; Dolui, K.; Datta, S.K. Patient Health Management System using e-Health Monitoring Architecture. In Proceedings of the 4th IEEE International Advance Computing Conference (IACC'14), New Delhi, India, 21–22 February 2014; IEEE Conference Publications: New Delhi, India, 2014; pp. 400–405. [CrossRef]
11. Patel, S.; Park, H.; Bonato, P.; Chan, L.; Rodgers, M. A review of wearable sensors and systems with application in rehabilitation. *J. NeuroEng. Rehabil.* **2012**, *9*. [CrossRef] [PubMed]

12. Khelil, A.; Shaikh, F.K.; Sheikh, A.A.; Felemban, E.; Bojan, H. DigiAID: A Wearable Health Platform for Automated Self-tagging in Emergency Cases. In Proceedings of the 4th International Conference on Wireless Mobile Communication and Healthcare (Mobihealth), Athens, Greece, 3–5 November 2014; IEEE Conference Publications: Athens, Greece, 2014; pp. 279–282. [CrossRef]

13. e-Health Sensor Platform V1.0 for Arduino and Raspberry PI from Cooking-Hacks. Available online: https://www.cooking-hacks.com/documentation/tutorials/ehealth-v1-biometric-sensor-platform-arduino-raspberry-pi-medical/ (accessed on 22 December 2018).

14. Feng, B.; Liu, B.; Pan, K. Vulture Voyeur a Sensor-Packed Egg Monitors Nests from the Inside. *IEEE Spectr.* **2016**, *53*, 17–18. [CrossRef]

15. Antonopoulos, C.; Panagiotakopoulos, T.; Panagiotou, C.; Touliatos, G.; Koubias, S.; Kameas, A.; Voros, N.S. On Developing a Novel Versatile Framework for Heterogeneous Home Monitoring WSN networks. *EAI End Trans. Pervasive Health Technol.* **2015**, *1*. [CrossRef]

16. Noueihed, J.; Diemer, R.; Chakraborty, S.; Biala, S. Comparing Bluetooth HDP and SPP for Mobile Health Devices. In Proceedings of the International Conference on Body Sensor Networks (BSN), Singapore, 7–9 June 2010; IEEE Conference Publications: Singapore, 2010; pp. 222–227. [CrossRef]

17. Granados, J.; Rahmani, A.M.; Nikander, P.; Liljeberg, P.; Tenhunen, H. Towards energy-efficient HealthCare: An Internet-of-Things architecture using intelligent gateways. In Proceedings of the International Conference on Wireless Mobile Communication and Healthcare (Mobihealth), Athens, Greece, 3–5 Novmber 2014; IEEE Conference Publications: Athens, Greece, 2014; pp. 279–282. [CrossRef]

18. MIThril, the Next Generation Research Platform for Context Aware Wearable Computing. Available online: https://www.media.mit.edu/wearables/mithril/ (accessed on 22 December 2018).

19. The Shimmer Platform. Available online: http://www.shimmersensing.com/ (accessed on 22 December 2018).

20. Samsung Simband. Available online: https://www.simband.io/ (accessed on 22 December 2018).

21. Ferreira, F.; Carvalho, V.; Soares, F.; Machado, J.; Pereira, F. Vital Signs Monitoring System Using Radio Frequency Communication: A Medical Care Terminal for Beddridden People Support. *Sens. Transducers* **2015**, *185*, 93–99. Available online: http://www.sensorsportal.com/HTML/DIGEST/february_2015/Vol_185/P_2607.pdf (accessed on 22 December 2018).

22. Srikanth, C.; Pradeep, D.S.; Charan, S. Smart Embedded Medical Diagnosis using Beaglebone Black and Arduino. *Int. J. Eng. Trends Technol.* **2014**, *8*, 43–48. [CrossRef]

23. Massot, B.; Baltenneck, N.; Gehin, C.; Dittmar, A.; McAdams, E. EmoSense: An Ambulatory Device for the Assessment of ANS Activity Application in the Objective Evaluation of Stress with the Blind. *IEEE Sens. J.* **2012**, *2*, 543–551. [CrossRef]

24. Ibrahim, H.; Ewais, S.; Chatterjee, S. A Novel, Low-Cost NeuroIS Prototype for Supporting Bio Signals Experimentation Based on BITalino. In *Information Systems and Neuroscience*; Gmunden Retreat on NeuroIS; Lecture Notes on Information Systems and Organization; Davis, F.D., Riedl, R., Brocke, J., Léger, P., Randolph, A., Eds.; Springer: London, UK, 2015; Volume 10, pp. 73–83. [CrossRef]

25. Da Silva, H.P.; Carreiras, C.; Lourenço, A.; Fred, A.; das Neves, R.C.; Ferreira, R. Off-the-person electrocardiography: Performance assessment and clinical correlation. *Health Technol.* **2015**, *4*, 309–318. [CrossRef]

26. Domingues, A. Smartphone Based Monitoring System for Long-Term Sleep Assessment. In *Springer Mobile Health Technologies Methods and Protocols*; Rasooly, A., Herold, K.E., Eds.; Methods in Molecular Biology; Springer: London, UK, 2015; Volume 1256, pp. 391–403. [CrossRef]

27. Verhulst, A.; Yamaguchi, T.; Richard, P. Physiological-based dynamic difficulty adaptation in a theragame for children with cerebral palsy. In Proceedings of the International Conference on Physiological Computing Systems (PhyCS), SCITEPRESS (Science and Technology Publications, Lda.), Angers, France, 11–13 February 2015; pp. 64–171. [CrossRef]

28. Fernandes, T.J.T.; Chęć, A.; Olczak, D.; Ferreira, H.A. Physiological computing gaming: Use of electrocardiogram as an input for video gaming. In Proceedings of the International Conference on Physiological Computing Systems (PhyCS), SCITEPRESS (Science and Technology Publications, Lda.), Angers, France, 11–13 February 2015; pp. 157–163. [CrossRef]

29. Birbas, M.; Petrellis, N.; Gioulekas, F. A Low Cost Sensor Controller for Health Monitoring. *J. Phys. Conf. Ser.* **2015**, *637*, 012021. [CrossRef]

30. Petrellis, N.; Birbas, M.; Gioulekas, F. Precision and Power Issues in a Medical Sensor Controller. In Proceedings of the 19th Panhellenic Conference on Informatics, (PCI'15), Athens, Greece, 1–3 October 2015; ACM Press: Athens, Greece, 2015; pp. 171–176. [CrossRef]

31. Petrellis, N.; Kosmadakis, I.E.; Vardakas, M.; Gioulekas, F.; Birbas, M.; Lalos, A. Compressing and Filtering Medical Data in a Low Cost Health Monitoring System. In Proceedings of the 21st Pan-Hellenic Conference on Informatics (PCI-2017), Larissa, Greece, 28–30 September 2017; ACM: New York, NY, USA, 2017; pp. 1–5, Article No. 40. [CrossRef]

32. Petrellis, N.; Birbas, M.; Vardakas, M.; Kosmadakis, I.E. Power saving issues in a low cost eHealth sensor controller. In Proceedings of the International Conference on Modern Circuits and Systems Technologies (MOCAST), Thessaloniki, Greece, 7–9 May 2018; pp. 1–4. [CrossRef]

33. Kosmadakis, I.E.; Petrellis, N.; Birbas, M.; Vardakas, M. Employing Savitzky-Golay Smoothing in a Low Cost eHealth Platform. In Proceedings of the International Conference on Telecommunications and Signal Processing (TSP), Athens, Greece, 4–6 July 2018; pp. 1–5. [CrossRef]

34. OASIS. MQTT 3.1.1 Protocol. Available online: http://mqtt.org/ (accessed on 22 December 2018).

35. Xu, H.; Caramanis, C.; Sanghavi, S. Robust PCA via outlier pursuit. *IEEE Tran. Inf. Theory* **2012**, *58*, 3047–3064. [CrossRef]

36. Carmi, A.; Gurfil, P.; Kanevsky, D. Methods for Sparse Signal Recovery Using Kalman Filtering With Embedded Pseudo-Measurement Norms and Quasi-Norms. *IEEE Trans. Signal Process.* **2010**, *58*, 2405–2409. [CrossRef]

37. Bosco, G. Principal Component Analysis of Electromyographic Signals: An Overview. *Open Rehabil. J.* **2010**, *3*, 127–131. [CrossRef]

38. Kanaujia, M.; Srivastava, G. ECG Signal Decomposition Using PCA and ICA. In Proceedings of the National Conference on Recent Advances in Electronics & Computer Engineering, Roorkee, India, 13–15 February 2015; pp. 301–305. [CrossRef]

39. Du, L.; Wang, B.; Wang, P.; Ma, Y.; Liu, H. Noise Reduction Method Based on Principal Component Analysis with Beta Process for Micro-Doppler Radar Signatures. *IEEE J. Sel. Top. Appl. Earth Obs. Remote Sens.* **2015**, *8*, 4028–4040. [CrossRef]

40. Hamadache, M.; Lee, D. Principal Component Analysis Based Signal-to-noise Ratio Improvement for Inchoate Faulty Signals: Application to Ball Bearing Fault Detection. *Int. J. Control Autom. Syst.* **2017**, *15*, 506–517. [CrossRef]

 *electronics*

Article

A Comparative Study of Markerless Systems Based on Color-Depth Cameras, Polymer Optical Fiber Curvature Sensors, and Inertial Measurement Units: Towards Increasing the Accuracy in Joint Angle Estimation

Nicolas Valencia-Jimenez [1,†], Arnaldo Leal-Junior [1,*,†], Leticia Avellar [1,†], Laura Vargas-Valencia [1], Pablo Caicedo-Rodríguez [2], Andrés A. Ramírez-Duque [1], Mariana Lyra [1], Carlos Marques [3], Teodiano Bastos [1] and Anselmo Frizera [1]

[1] Graduate Program in Electrical Engineering, Universidade Federal do Espírito Santo, 29075-910 Vitoria, Brazil; valenciaj@ieee.org (N.V.-J.); leticia.avellar@aluno.ufes.br (L.A.); laura.valencia@aluno.ufes.br (L.V.-V.); andres.duque@aluno.ufes.br (A.A.R.-D.); mariana.silveira@aluno.ufes.br (M.L.); teodiano.bastos@ufes.br (T.B.); frizera@ieee.org (A.F.)

[2] Grupo de Investigación en Tecnologías y Ambiente, Corporación Universitaria Autónoma del Cauca, 190001-19 Popayán, Colombia; pablo.caicedo.r@uniautonoma.edu.co

[3] Instituto de Telecomunicações and I3N & Physics Department, University of Aveiro, Campus Universitário de Santiago, 3810-193 Aveiro, Portugal; carlos.marques@ua.pt

* Correspondence: leal-junior.arnaldo@ieee.org

† These authors contributed equally to this work.

Received: 15 December 2018; Accepted: 30 January 2019; Published: 2 February 2019

Abstract: This paper presents a comparison between a multiple red green blue-depth (RGB-D) vision system, an intensity variation-based polymer optical fiber (POF) sensor, and inertial measurement units (IMUs) for human joint angle estimation and movement analysis. This systematic comparison aims to study the trade-off between the non-invasive feature of a vision system and its accuracy with wearable technologies for joint angle measurements. The multiple RGB-D vision system is composed of two camera-based sensors, in which a sensor fusion algorithm is employed to mitigate occlusion and out-range issues commonly reported in such systems. Two wearable sensors were employed for the comparison of angle estimation: (i) a POF curvature sensor to measure 1-DOF angle; and (ii) a commercially available IMUs MTw Awinda from Xsens. A protocol to evaluate elbow joints of 11 healthy volunteers was implemented and the comparison of the three systems was presented using the correlation coefficient and the root mean squared error (RMSE). Moreover, a novel approach for angle correction of markerless camera-based systems is proposed here to minimize the errors on the sagittal plane. Results show a correlation coefficient up to 0.99 between the sensors with a RMSE of 4.90°, which represents a two-fold reduction when compared with the uncompensated results (10.42°). Thus, the RGB-D system with the proposed technique is an attractive non-invasive and low-cost option for joint angle assessment. The authors envisage the proposed vision system as a valuable tool for the development of game-based interactive environments and for assistance of healthcare professionals on the generation of functional parameters during motion analysis in physical training and therapy.

Keywords: joint angular kinematics; human motion analysis; RGB-D cameras; polymer optical fiber; inertial measurement units

---

Electronics **2019**, *8*, 173; doi:10.3390/electronics8020173   www.mdpi.com/journal/electronics

1. Introduction

There is a clear and growing interest in developing technological-based tools that systematically analyze human movement. Notably, there are many advantages to implement automated systems to detect human motion for applications associated with children in a healthcare context or to assess mobility impairment of ill and elderly people [1]. Automated quantification of body motion to support specialists in the decision-making process such as stability, duration, coordination, and posture control is the desired result for those technological-based approaches [2,3]. Despite recent advances in this area, automated quantification of the human movement for children with sensory processing and cognitive impairments, and adults with mobility disabilities presents multiple challenges due to factors such as accessibility barriers, attached to the body requirements, and the high cost of the system.

Automated analysis of body movements typically involves obtaining 3D joint data such as position and orientation, which are estimated in two different ways using an intrusive or a non-intrusive approach, also known as wearable and non-wearable technologies [4]. Wearable systems are portable and can be used by people with movement impairments in unstructured scenarios [5]. However, advances in non-wearable sensing technologies and processing techniques have appeared to measure with high accuracy human biomechanics in highly structured environments [6]. For instance, camera-based markerless systems can be used in scenarios when the user does not admit a wearable device to capture data [4].

Commonly, analysis of joint angles is conducted through portable wearable sensors, which include electrogoniometers and potentiometers mounted in a single axis; however, they are bulky and limit natural patterns of movement [7]. Flexible goniometers adapt better to body parts and are not sensitive to misalignments due to movements of polycentric joints. In addition, they employ strain gauges, which must be carefully attached to the skin due to their high sensitivity [8]. Advancements in micro-electro-mechanical systems (MEMs) have enabled the growth of new wearable sensors.

Despite the wide range of applications, inertial measurement units (IMUs) present high sensitivity to the magnetic field and need frequent calibration [9]. New sensor technologies, such as optical sensors (with advantages such as compactness, lightweight, flexibility, and insensitivity to electromagnetic interference [10]), are employed for joint angle measurements using different approaches. Those approaches include intensity variation [11], fiber Bragg gratings [12] and interferometers [13]. Considering their advantages such as low cost, simplicity on the signal processing, and portability [14], intensity variation-based sensors emerge as a feasible alternative for joint angle assessment in movement analysis and robotic applications [15]. However, to date, POF-based curvature sensors share the same limitations of the aforementioned goniometers, since angle assessment is limited to a single plane.

Non-wearable human motion analysis systems use either multiple image sensors or multiple motion sensors [16]. The latter can be classified as either marker-based or markerless capture [17]. Although commercial marker-based systems can track human motion with high accuracy, markerless systems have many advantages [18]. Markerless systems can eliminate the difficulty of applying markers to users with physical or cognitive limitations [19].

However, most of the markerless systems require a surface model. Therefore, 3D surface reconstruction is a prerequisite to markerless capture [17,20]. The advent of off-the-shelf depth sensors, such as Microsoft Kinect [21], makes it easy to acquire depth data. Such depth data are beneficial for gesture recognition [22]. However, motion capture with depth sensors is a challenging issue due to the limitations of accuracy and range [23].

To build an efficient system, it is necessary to identify its functionalities and requirements to effectively contribute to the user's necessities [24]. For example, people with disabilities that harm the use of wearable sensors impose difficulties on the acquisition of mobility parameters by clinical staff. Consequently, it is necessary to propose valid and straightforward methods for acquiring such parameters.

The main assumption of the proposed technique for accuracy enhancement is the correlation between the measurement errors with the anthropometric parameters of the user. Thus, a compensation equation was obtained, which correlates the errors with the user's anthropometric parameters. Although the proposed approach requires a longer calibration step prior to the sensor system application, it results in lower errors when compared to conventional approaches for angle analysis using markerless systems. The novel approach proposed here is capable of enhancing the accuracy of non-wearable systems. In addition, it is important to emphasize that such an approach can be applied in different systems (even the more accurate ones) for the enhancement of their accuracy. Furthermore, this technique can, with a few adjustments on the calibration step, be applied on the assessment of 3D dynamics.

This paper presents the analysis and comparison among a multi-red green blue-depth (RGB-D) vision system, an intensity variation-based POF sensor, and IMUs, which are intended for human joint elbow angles assessment. These wearable sensors were chosen due to their compactness, which would not cause occlusions for the markerless camera system. In addition, a marker-based camera system was not used in this comparison, since the markers can aid or harm the joint detection by the markerless system, which would not represent the condition for a practical application. This systematic comparison aims to study the trade-off between the markerless feature of the vision system and its accuracy by comparing it with wearable technologies for joint angle measurements. Thus, a compensation technique based on anthropomorphic measurements of the participants was proposed and validated using the POF curvature sensor measures.

2. Materials and Methods

2.1. RGB-D Fusion System

This system was designed as a distributed and modular architecture using the open source project Robot Operating System (ROS). The architecture developed here was built using a node graph approach. This system consists of several nodes to local video processing, distributed around several different hosts and connected at runtime in a peer-to-peer topology. The inter-node connection is implemented as a hand-shaking and occurs in XML-RPC protocol (Remote Procedure Call protocol which uses XML to encode its calls). The node structure is flexible, scalable, and can be dynamically modified, i.e., each node can be started and left running along an experimental session or resumed and connected to each other at runtime.

A computer vision framework composed of an unstructured and scalable network of RGB-D cameras is used to automatically estimate joint position, this visual sensor network counteracts typical problem as occlusion and narrow field of view. Consequently, the system uses a distributed architecture for processing the videos of each sensor independently. In this structure, a human body analysis algorithm is executed for each camera. Subsequently, the data from each sensor are transformed and represented respect to a shared reference to fusion them and generate the global joint position.

The vision system is composed of two RGB-D cameras, as shown in Figure 1, which each camera is connected to a workstation equipped with a processor Intel Core i5 and a GeForce GTX GPU board (GTX960 board, and GTX580 board) to execute local data processing. All workstations are connected through a local area network synchronized using Network Time Protocol (NTP) and managed through ROS. The data fusion estimation is sent for a third-party software interaction, which are executed on the highest processing capacity workstation.

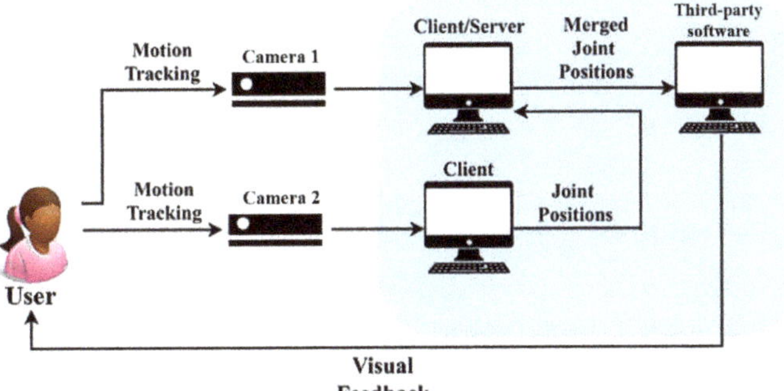

Figure 1. Configuration of the RGB-D system.

Each workstation has two primary processes: user detection and position/orientation estimation of 15 joints. The detection process targets the candidate points to be people in the scenario, using the software NiTE, which does the client task, sending the movement estimation to the server over the network. The extrinsic (transformation from 3-D world coordinate system to the 3-D camera's coordinate system) and intrinsic (transformation from the 3-D camera's coordinates into the 2-D image coordinates) calibration of each RGB-D camera is performed using both the OpenCV package and the multi-camera network calibration tool provided by OpenPTrack (see Figure 2).

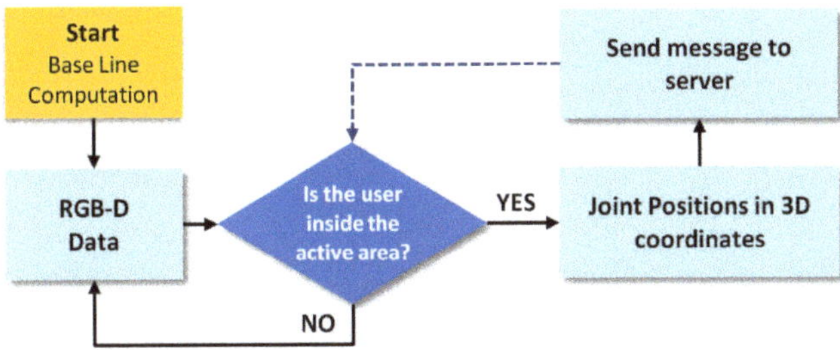

Figure 2. Client flowchart.

The workstation with the highest processing capacity is also used as the system server, which is responsible for the vision fusion process, using a Kalman filter. When the server receives a message with the position data of a client, it checks the time interval between the last received message and the current message. If this interval is greater than 33 ms, the system discards the received measurement and resumes counting the intervals from the next measure to receive. This is made to do not use the merging data with a very discrepant time. If the data is within the time interval, the system transforms the client coordinate system into the global coordinate system defined in the extrinsic calibration process. Then, the data is inserted into the Kalman filter. The saved data is processed through a low-pass Butterworth filter used to eliminate noise and to achieve a smoother estimate (see Figure 3).

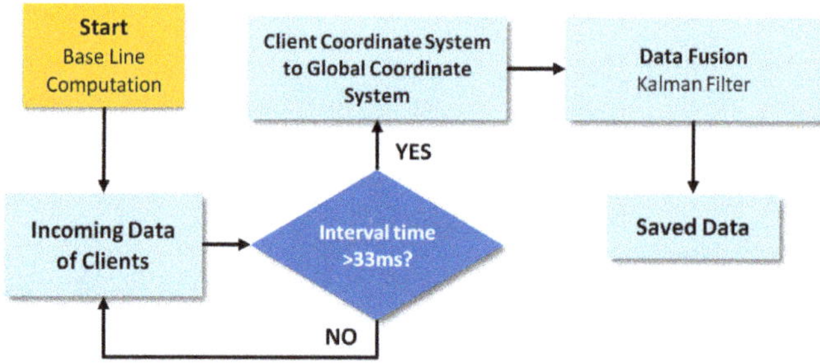

Figure 3. Server flowchart.

The RGB-D fusion process is executed to obtain kinematic parameters and corporal patterns to be shown in an assessment interface that detects and quantifies the user's movements. The system can produce parameters such as range of motion and positions of the tracked body articulations in three dimensions. In the same way, the system can be configured to show specific articulation angles. We use the law of cosine to calculate the elbow angle, as suggested by different authors [25–28]. Equation (1) shows the relation between the forearm d_1 and upper arm d_2 lengths as shown in Figure 4. The blue dots shown in Figure 4 are identified using the software NiTE, The blue dots shown in Figure 4 are identified using the software NiTE, such as aforementioned, where three points are identified (on the shoulder, elbow and hand), with (X, Y, Z) coordinates represented at each point.

$$\theta = \cos^{-1}\left(\frac{-d_3 + d_1 + d_2}{2 * d_1 * d_2}\right) \tag{1}$$

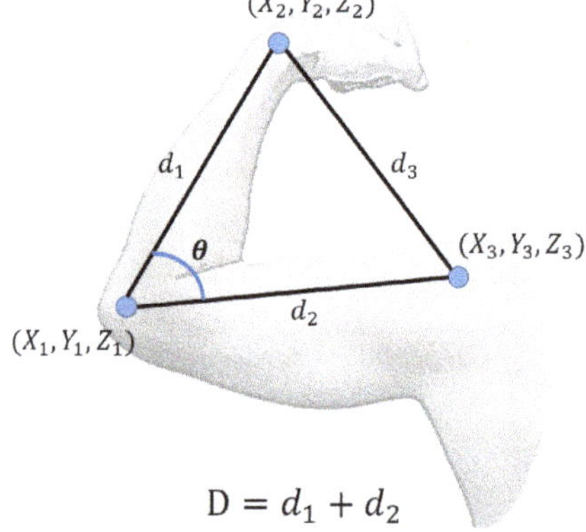

Figure 4. Parameters to calculate the articulation angle of any joint.

2.2. POF Sensor

The POF curvature sensor used in this work is based on the intensity variation principle in which, considering the fiber input connected to a light source and the output to a photodetector, there is a power variation on the POF output when it is subjected to curvature. Such power variation is proportional to the curvature angle and occurs due to some effects on the fiber such as radiation losses, light scattering, and stress-optic effects, as thoroughly discussed in Leal-Junior et al. [29]. To increase the sensor linearity and sensitivity as well as the hysteresis reduction, a lateral section is made on the fiber, removing the fiber cladding and part of its core (considering the core-cladding structure of conventional solid-core optical fibers [30]). This material removal on the fiber creates a so-called sensitive zone, where the fiber is more sensitive to curvature variations.

The POF employed in this work, multimode HFBR-EUS100Z POF (Broadcom Limited, Singapore), has its core made of Polymethyl Methacrylate (PMMA) with 980 µm, whereas the cladding thickness is 10 µm. In addition, the fiber also has a polyethylene coating for mechanical protection, which results in a total diameter of 2.2 mm for this fiber. The lateral section is made on the fiber through abrasive removal of material, where the depth and length of the sensitive zone are about 14 mm and 0.6 mm, respectively. The sensitive zone length and depth were chosen as the ones that result in high sensitivity, linearity with low hysteresis, as experimentally demonstrated in Leal-Junior et al. [31]. The POF with the lateral section has one end connected to a light emitting diode (LED) IF-E97 (Industrial Fiber Optics, Tempe, AZ, USA) with central wavelength at 660 nm, whereas the other end (output) is connected to the photodiode IF-D91 (Industrial Fiber Optics, Tempe, AZ, USA). The POF curvature sensor needs a characterization step prior to its application in movement analysis, where this characterization is performed by positioned the POF sensor on the experimental setup shown in Figure 5. In this setup, a DC motor with controlled position and angular velocity performs flexion and extension movements on the POF sensitive zone on an angular range of 0–90°. Then, the power attenuation is compared with the angle measured by the potentiometer (also shown in Figure 5) and a linear regression is made.

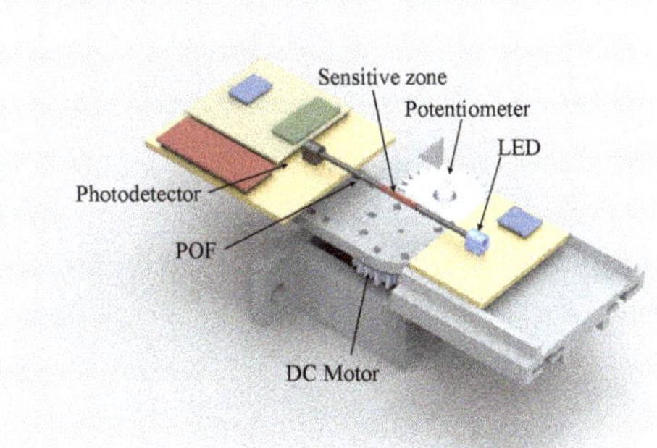

Figure 5. Experimental setup for POF curvature sensor characterization.

To get user movement parameters, the sensor is positioned on their elbow joints. Since the mounting conditions of the POF curvature sensor has direct influence on its response (as discussed in Leal-Junior et al. [31]), each user is asked to perform a 90° angle in the sagittal plane (in both cases, i.e., sensor at elbow and knee joints). Thus, the 90° movement on the sagittal plane is used to calibrate the sensor as a function of its positioning on each subject.

2.3. IMU

Two wireless MTw IMU sensors (Xsens, Enschede, The Netherlands) are used, which are placed on the users' body segments, (weight: 10 g, size: 47 × 30 × 13 mm). Each MTw sensor comprises a 3D gyroscope, a 3D accelerometer and a 3D magnetometer. IMUs sends its data to a station (Awinda, Xsens) via a proprietary wireless communication protocol at 100 Hz of sampling frequency. The orientation data, in quaternion format, was estimated with a customized Kalman filter developed by the manufacturer (XKF3-hm) [32].

We implemented a method using two IMU units to estimate the elbow flexion-extension angle. The IMU on the proximal body limb (i.e., upper arm) was used as a reference, hereafter referred to as S_1. The second IMU, which was located on the distal body limb, will be referred as S_2. The algorithm uses five coordinate systems for its operation. S_1 and S_2 are the local coordinate system of the IMU sensors, these are embedded in each sensor. The technical-anatomical coordinate systems (B_1 and B_2) correspond to the body limbs' orientation [33]. Finally, the global coordinate system, G, is a frame formed by gravity acceleration vector and Earth's magnetic north. The reference IMU sensor (S_1) was carefully aligned with the body limb, in such a way, during calibration moment, with a neutral standing posture and after applying a gravity alignment, B_1 (B_2) coordinate system is estimated from S_1 (S_2).

The algorithm aims the estimation of the angle between the orientation of the proximal and distal limb. Therefore, the limbs' orientations should be projected to a common coordinate system (i.e., global coordinate system). This is done through the orientation quaternions, ${}^G\mathbf{q}_{B_1}(t)$ and ${}^G\mathbf{q}_{B_2}(t)$. The estimation process presupposes that x-axis of both technical-anatomical coordinate system are aligned with gravity vector at the calibration moment in a straight neutral posture, as proposed in previous works [33–35]. Besides the elbow joint is simplified as a 1-DOF hinge joint. The flexion-extension angle (α) is calculated following the steps below:

- To calculate the rotation quaternion, $\mathbf{q}_c$, that aligns the x-axis of the S_1 coordinate system, $\mathbf{x}_{S_1}$, with the gravity vector is expressed in the global coordinate system ($\mathbf{z}_G$) at the calibration moment, as proposed in [33].
- To define the technical-anatomical coordinate systems ${}^G\mathbf{q}_{B_1}(0)$ and ${}^G\mathbf{q}_{B_2}(0)$, at the calibration moment, applying Equations (2) and (3).

$$ {}^G\mathbf{q}_{B_1}(0) = \mathbf{q}_c \otimes {}^G\bar{\mathbf{q}}_{S_1}(0) \tag{2} $$

$$ {}^G\mathbf{q}_{B_1}(0) = {}^G\mathbf{q}_{B_2}(0) \tag{3} $$

where $\otimes$ denotes the Hamilton product. Please note that due to the assumptions mentioned above, at the calibration moment, ${}^G\mathbf{q}_{B_1}(0)$ and ${}^G\mathbf{q}_{B_2}(0)$ are equals. Which means that the initial angle in a straight neutral posture is zero.

- To calculate sensor orientation S_1 (S_2) with respect to its associated body segment B_1 (B_2) using Equation (4). Please note that ${}^{B_1}\mathbf{q}_{S_1}$ and ${}^{B_2}\mathbf{q}_{S_2}$ are constant at all time instants.

$$ {}^{B_1}\mathbf{q}_{S_1} = \left({}^G\mathbf{q}_{B_1}(0)\right)^{-1} \otimes {}^G\bar{\mathbf{q}}_{S_1}(0) \tag{4} $$

- To estimate the body segments' orientation ${}^G\mathbf{q}_{B_1}(t)$ and ${}^G\mathbf{q}_{B_2}(t)$ using Equation (5). Please note that Equations (4) and (5) are expressed in terms of B_1 and S_1, but they are also applicable to B_2 and S_2.

$$ {}^G\mathbf{q}_{B_1}(t) = {}^G\mathbf{q}_{S_1}(t) \otimes \left({}^{B_1}\mathbf{q}_{S_1}\right)^{-1} \tag{5} $$

- To calculate the relative orientation between ${}^G\mathbf{q}_{B_1}(t)$ and ${}^G\mathbf{q}_{B_2}(t)$. In that way, ${}^{B_1}\mathbf{q}_{B_2}(t)$ is then decomposed into 'ZXY' Euler angles. The rotation about z-axis represents the elbow flexion-extension angle, α, consistent with the ISB recommendations [16,36].

$$ {}^{B_1}\mathbf{q}_{B_2}(t) = \left({}^G\mathbf{q}_{B_1}(t)\right)^{-1} \otimes {}^G\mathbf{q}_{B_2}(t) \tag{6} $$

2.4. Sensors Characterization

To validate the measurements of the camera-based and IMU-based systems a goniometer as a standard reference is used. This reference has two adjustable lever locks, which are positioned to limit the flexion-extension motion between 20° (lower bound) and 90° (upper bound). The goniometer was placed and aligned with the elbow joint of the subject M1, then, this was asked to perform flexion-extension movements on the sagittal plane in such a way to reach both locks. Lastly, the data was acquired by the camera-based system and the IMUs. The POF curvature sensor is characterized using the procedure mentioned in Section 2.2.

2.5. Experimental Protocol

Eleven participants without motor impairments were enrolled in this study. Six females, referred as F1, F2, F3, F4, F5, and F6, age: 27.3 ± 4.9 years old, corporal weight 56.8 ± 16.3 kg, and five males, referred as M1, M2, M3, M4, and M5, age: 27.4 ± 3.3 years old, corporal weight 70.2 ± 3.8 kg, as shown in Table 1. This research was approved by the Ethical Committee of UFES (Research Project CAAE: 64797816.7.0000.5542). As shown in Table 1, there is a variability not only in gender, but also on subjects' weight and height, which provides evidence for a further generalization of the proposed study.

Table 1. Characteristics of the participants of this research.

Subject	Age (Years)	Height (cm)	Weight (Kg)
M1	30	163	67
M2	22	176	66
M3	30	173	73
M4	27	183	75
M5	28	170	70
F1	30	156	57
F2	32	158	46
F3	22	160	48
F4	23	163	53
F5	24	158	48
F6	33	176	89

Two IMU sensors, one POF curvature sensor and two RGB-D cameras were used to estimate elbow joint angles. The IMU reference sensor was placed on the superior third of the right upper arm, and the second IMU was attached dorso-distally on the right forearm, as shown in Figure 6e. In a standing neutral posture, both sensors were positioned with x-axis pointing cranially, z-axis laterally and y-axis an orthogonal axis to x and z axes. These positions have been suggested by different authors [9,35,37]. Moreover, the POF curvature sensor was carefully aligned with the elbow joint in such a way that the sensitive zone of the optical fiber is located on the axis of rotation (flexion-extension axis).

In this test, the 11 participants (see Table 1) perform, a comfortable self-velocity, flexion-extension movements on three planes: sagittal, transverse, and frontal. Each participant was standing at the center of the room, observing the middle point between the two RGB-D cameras, see Figure 6d. All trials start with a synchronization movement, which consists of keeping the elbow in maximum extension on standing posture, then performing an elbow flexion of 90° and returning to the extended elbow position, where each transition last 5 s. Then, the subject was asked to perform three repetitions of flexion-extension on a specific plane. In the sagittal plane, the shoulder is in a neutral position and the participant performs elbow flexion-extension to get the maximum angle as possible (see in Figure 6a). In the transverse plane, the shoulder is in abduction (at max 90°) and kept in that position for 5 s before the elbow flexion-extension movements (Figure 6b). In the frontal plane, the shoulder is in abduction (at max 90°) and external rotation, so the palm of the hand is facing forward as shown in Figure 6c. These described steps can be summarized in Figure 7.

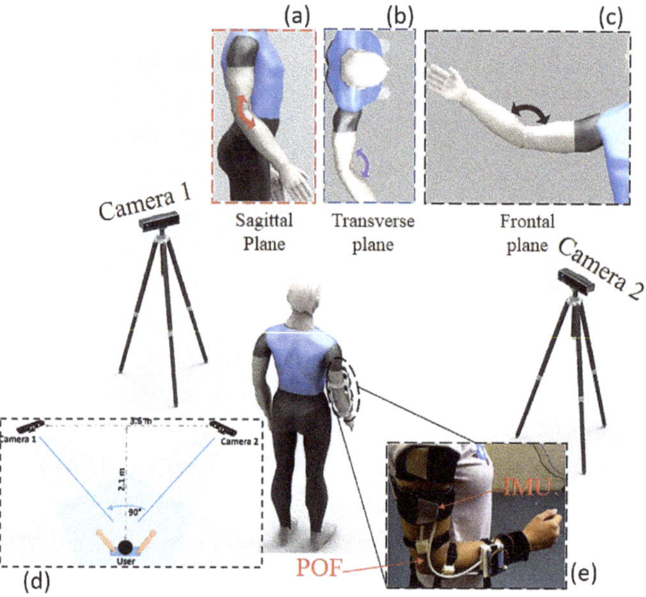

Figure 6. Sensors' placement on the human upper limb, and movements performed during the experimental protocol. (**a**) User movement representation in sagittal plane, (**b**) transverse plane, and (**c**) frontal plane. (**d**) Top view of RGB-D system arrangement. (**e**) User using the IMU system and POF sensor.

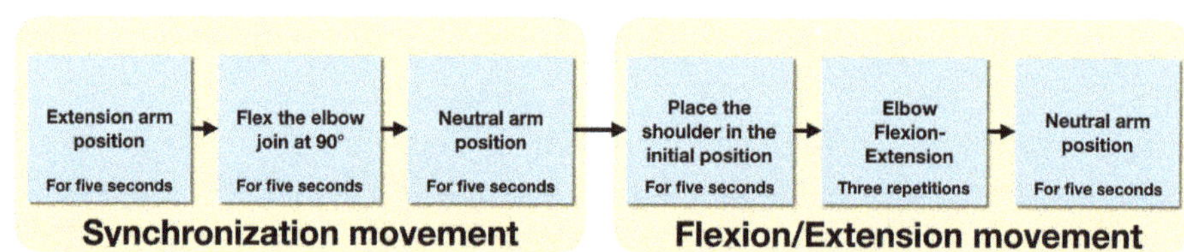

Figure 7. Summary of the protocol's phases.

3. Results and Discussion

The comparison variables of the three systems were: (i) the correlation coefficient and (ii) the root mean squared error (RMSE) between the RGB-D cameras, IMUs, and POF. After the comparison among the sensors, a novel approach for angle correction on the markerless camera-based system was proposed and validated for the correction of angular errors on the sagittal plane. The proposed technique is based on anthropometric measurements of each subject, where the errors of the camera-based system compared with the POF curvature sensor and IMU system are corrected through a correlation between the measured errors and the length of the arm of each subject (d_1, d_2 and d_3 of Equation (1)). In this way, an equation for error correction considering the length d_3 of each subject was obtained.

3.1. Sensors Characterization

3.1.1. IMUs and RGB-D Cameras

Figure 8 presents the results for RGB-D camera system and IMUs for the characterization phase using a goniometer as a reference for upper and lower bounds. The performed movement was divided into three cycles, and each cycle presents correlation between cameras and IMUs higher than 0.98.

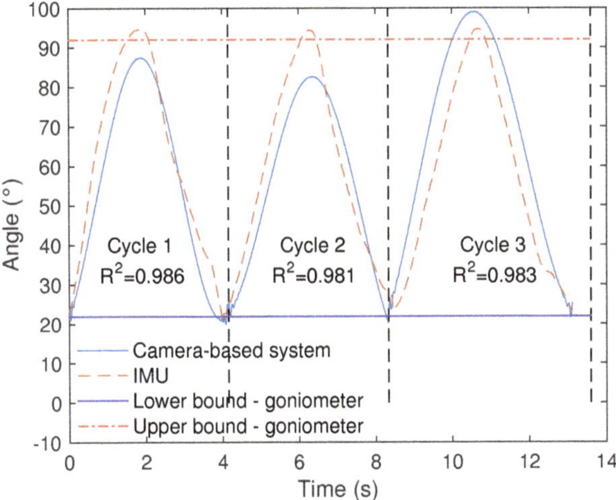

Figure 8. Camera-based system and IMU angles of the first test.

Table 2 shows the maximum and minimum angles for cameras system and IMUs, compared to upper and lower bounds, respectively. The average error for the cameras system was 4.9°, with a maximum error of 9°, when compared with the goniometer for two values (90° e 20°), which is lower than the mean error presented in Tannous et al. [38] (14, 6°). However, in Tannous et al. [38] only one camera was used, carrying a higher self-occlusion leading to errors on the angle assessment. Since our system consists of two cameras, the self-occlusion decreases, consequently, reducing the errors. Compared to the goniometer, the IMUs average error was 3.7°, which is lower than the camera-based system.

Table 2. Maximum and minimum angles of camera-based system and IMU of each cycle for the first test.

	Cycle 1		Cycle 2		Cycle 3	
	Max [°]	**Min [°]**	**Max [°]**	**Min [°]**	**Max [°]**	**Min [°]**
Camera	87.4	21.0	82.6	12.8	99.0	22.2
IMU	94.6	22.9	94.4	22.7	94.7	23.0
Goniometer	90.0	20.0	90.0	20.0	90.0	20.0

3.1.2. POF

Figure 9 presents the normalized POF power attenuation as a function of the angle on the experimental setup shown in Figure 6, where a high correlation with the measured angle was obtained, since the determination coefficient (R^2) between the POF power variation and the angle was 0.996.

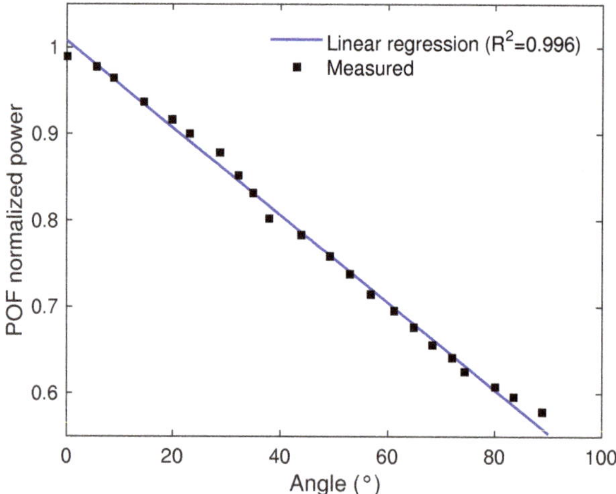

Figure 9. Sensors' placement on the human upper limb and the movements performed.

Thereafter, the POF curvature sensor is positioned on the subject's elbow joint as previously described (see Section 2.5). The data of the elbow movement at each plane (sagittal, frontal and transverse) were recorded, which are presented in Figure 10 for subject M3. The initial movement of 90° in the sagittal plane was used to adjust the sensor response with respect to the differences aroused from the positioning on the subject elbow.

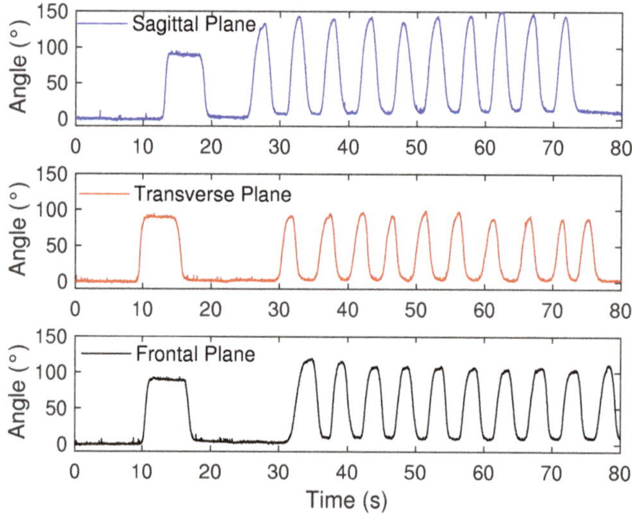

Figure 10. Sensors' placement on the human upper limb and the movements performed.

The sensor response has the pattern of the flexion/extension movements performed by the subject. Furthermore, the angles estimated by the POF curvature sensor are within the range of movement described by the literature for elbow (0–145°) [39]. Thus, the results presented in Figure 10 in conjunction with the high correlation of 0.996 (obtained on the sensor characterization as a function of the angle measured by a potentiometer) shows the feasibility of the POF sensor on the angle assessment, which makes it a suitable option for the comparison with angles estimated by the camera-based system.

3.2. Comparison among the Sensors

After the first evaluation of the sensors and the POF curvature sensor characterization, the sensors used in this work were compared using the experimental protocol depicted in Section 2.5. In this case, the camera-based system was compared with both IMU and POF. The comparison was made with respect to the correlation coefficient and RMSE (as also performed in the previous sections). Figure 11 shows the results obtained for all sensors in different planes, i.e., sagittal, transverse, and frontal planes, for subject M1.

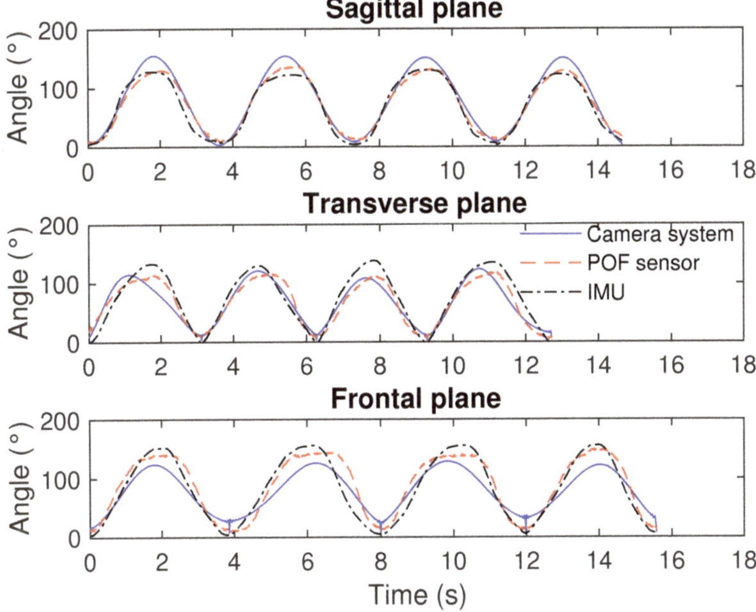

Figure 11. Comparison among camera-based system, POF curvature sensor and IMU in the sagittal, frontal, and transverse planes, for subject M1.

The results presented in Figure 11 show a good correlation between the errors of the POF curvature sensor and IMU, especially on sagittal and frontal planes. Although we used the same number of cycles to compare the sensors, the period of each movement is different, due to the fact that each subject was allowed to perform the movements at a comfortable self-velocity.

Furthermore, the range of movement at each plane is different, i.e., the movement at the sagittal plane occurs in a range of about 0–145°, whereas the one at the transverse plane reaches angles lower than

130°. Similarly, the angles at the frontal plane can be as high as 145° (as in the sagittal plane). From the tests, the mean deviation between POF curvature sensor and IMUs was about 6.5% on the tests in the sagittal plane. However, such deviation increased to about 10% on the transverse and frontal planes. The reason for this increase can be related to the POF positioning on the tests, since it is a critical factor on the angle assessment using such technology. In addition, it can be also related to the increase of the errors of the IMUs when the test was performed in planes different from the sagittal one, as reported in Vargas-Valencia et al. [33]. Regarding the camera-based system, the results at the sagittal plane show an overestimation of the angle, when compared to IMU and POF curvature sensor. In this case, the angles estimated by the markerless camera system had a maximum value of about 160°, which is higher than the elbow range of motion [39]. In contrast, the camera-based system underestimates the angles at the frontal plane when compared with the other two systems for angle assessment.

Such as aforementioned, the errors on the markerless camera system for angle assessment are related to issues, such as frame errors, exploitation of multiple image streams and, especially, due to self-occlusions. To further evaluate the errors obtained by the camera-based system, Table 3 presents the correlation coefficient and RMSE between the markerless system and the IMUs for each of the 11 participants in all three planes tested, whereas Table 4 presents the correlation and RMSE between the markerless system and the POF curvature sensor.

Table 3. RMSE and correlation coefficient for Cameras and IMU.

	Sagittal		Frontal		Transverse	
	R^2	RMSE	R^2	RMSE	R^2	RMSE
M1	0.994 ± 0.0041	$13.53° \pm 2.87°$	0.955 ± 0.0164	$14.31° \pm 6.37°$	0.988 ± 0.0043	$18.62° \pm 1.32°$
M2	0.994 ± 0.0023	$10.21° \pm 1.86°$	0.983 ± 0.0213	$11.96° \pm 4.67°$	0.991 ± 0.0052	$14.48° \pm 4.53°$
M3	0.986 ± 0.0052	$7.42° \pm 2.51°$	0.993 ± 0.0063	$15.03° \pm 1.55°$	0.984 ± 0.0060	$18.84° \pm 7.23°$
M4	0.984 ± 0.0058	$13.02° \pm 3.12°$	0.982 ± 0.0026	$9.99° \pm 3.94°$	0.992 ± 0.0031	$15.89° \pm 6.59°$
M5	0.991 ± 0.0025	$9.60° \pm 3.08°$	0.989 ± 0.0095	$14.34° \pm 3.85°$	0.978 ± 0.0230	$11.21° \pm 2.90°$
F1	0.993 ± 0.0011	$11.83° \pm 2.93°$	0.976 ± 0.0098	$16.42° \pm 2.41°$	0.991 ± 0.0077	$17.61° \pm 3.67°$
F2	0.990 ± 0.0043	$11.26° \pm 2.89°$	0.986 ± 0.0045	$8.49° \pm 5.61°$	0.921 ± 0.0833	$15.28° \pm 4.31°$
F3	0.974 ± 0.0136	$12.04° \pm 5.39°$	0.990 ± 0.0046	$14.89° \pm 1.90°$	0.956 ± 0.0543	$16.73° \pm 2.89°$
F4	0.996 ± 0.0019	$11.90° \pm 3.53°$	0.989 ± 0.0057	$15.98° \pm 6.78°$	0.990 ± 0.0052	$16.29° \pm 5.98°$
F5	0.993 ± 0.0079	$9.89° \pm 0.77°$	0.994 ± 0.0020	$12.56° \pm 3.74°$	0.987 ± 0.0037	$19.42° \pm 5.03°$
F6	0.995 ± 0.0042	$11.81° \pm 0.52°$	0.989 ± 0.0034	$17.31° \pm 4.04°$	0.992 ± 0.0006	$19.98° \pm 6.52°$

Table 4. RMSE and correlation coefficient for Cameras and POF.

	Sagittal		Frontal		Transverse	
	R^2	RMSE	R^2	RMSE	R^2	RMSE
M1	0.988 ± 0.0066	$10.01° \pm 2.12°$	0.972 ± 0.0175	$15.32° \pm 6.92°$	0.994 ± 0.0033	$19.32° \pm 3.30°$
M2	0.977 ± 0.0055	$10.90° \pm 1.67°$	0.985 ± 0.0065	$11.28° \pm 0.97°$	0.985 ± 0.0080	$14.98° \pm 4.83°$
M3	0.978 ± 0.0110	$6.90° \pm 2.25°$	0.967 ± 0.0025	$16.14° \pm 2.74°$	0.981 ± 0.0123	$19.83° \pm 3.80°$
M4	0.955 ± 0.0208	$12.00° \pm 2.83°$	0.916 ± 0.0106	$13.35° \pm 4.23°$	0.965 ± 0.0165	$13.53° \pm 5.28°$
M5	0.994 ± 0.0017	$10.87° \pm 0.15°$	0.973 ± 0.0069	$13.26° \pm 3.39°$	0.954 ± 0.0289	$11.60° \pm 1.17°$
F1	0.975 ± 0.0178	$11.52° \pm 3.24°$	0.978 ± 0.0196	$14.84° \pm 1.09°$	0.970 ± 0.0205	$17.84° \pm 3.44°$
F2	0.983 ± 0.0120	$10.35° \pm 0.60°$	0.987 ± 0.0051	$9.42° \pm 4.27°$	0.945 ± 0.0375	$17.44° \pm 4.77°$
F3	0.982 ± 0.0085	$7.72° \pm 1.59°$	0.978 ± 0.0070	$17.17° \pm 1.80°$	0.982 ± 0.0083	$17.86° \pm 5.20°$
F4	0.978 ± 0.0098	$12.76° \pm 1.63°$	0.975 ± 0.0076	$16.95° \pm 4.46°$	0.987 ± 0.0001	$17.71° \pm 6.52°$
F5	0.989 ± 0.0097	$8.11° \pm 0.81°$	0.977 ± 0.0070	$13.56° \pm 4.30°$	0.979 ± 0.0131	$19.28° \pm 5.49°$
F6	0.964 ± 0.0076	$13.52° \pm 0.95°$	0.892 ± 0.0165	$19.13° \pm 4.53°$	0.979 ± 0.0047	$18.92° \pm 5.36°$

Tables 3 and 4 show a correlation coefficient higher than 0.9 in all analyzed cases, which indicates a high correlation between the responses of the sensors. In addition, the standard deviation of the correlation coefficient was below 0.01 in all the analyzed cases. Thus, it is possible to verify not only a high correlation between the data of the camera-based system compared to the wearable ones, but also that the results present a promising evidence of repeatability of such systems. The mean of correlation coefficients between the camera-based system and the IMUs were 0.990, 0.984 and 0.979 on the sagittal, transverse, and frontal planes, respectively. It is noteworthy that higher correlations were obtained between the camera-based system and the IMUs than the ones comparing the markerless system with the POF curvature sensor. The mean of the correlation coefficients considering the later comparison was 0.978, 0.964 and 0.975 for sagittal, transverse, and frontal planes, respectively.

Even though the proposed camera-based system presented high correlation with the wearable sensors in all scenarios, the errors of such system are generally high. Such as can be observed in Figure 11, there are deviations on the angle estimation of the camera-based system when compared with the wearable sensors, where, considering all the performed tests, these errors can be as high as 15° on the worst case. In addition, the mean errors are about 10° when compared with the wearable sensors. It is noteworthy that these errors are lower than the ones reported on the literature [40], which is mainly due to the use of two cameras to reduce the errors related to occlusions. However, errors of about 10° are still not sufficient when a reliable system for movement analysis is concerned. Nevertheless, the high correlations obtained in all tests for the comparison with the wearable sensors (see Tables 3 and 4) indicate that the proposed markerless camera-based system can be a feasible solution for angle estimation if a post-processing technique for the correction of the angular errors is applied, as also discussed in Schmitz et al. [40].

3.3. Technique for Angle Correction in Markerless Camera-Based Systems

The primary assumption for the proposed compensation technique for angle errors in markerless camera-based systems is that the errors mainly occur due to occlusions, or errors on computer vision algorithm for the tracking of the anatomical points used to calculate the parameters d_1 and d_2 in Figure 4. If these parameters are incorrectly estimated, errors on the angle assessment will occur. Thus, it is possible to assume that these angular errors have a correlation with the anthropometric measurements of each participant. To verify this assumption and develop the compensation technique for the markerless system, each participant performed 3 flexion/extension cycles only on the sagittal plane (see Section 2.5), and the angles estimated with the markerless camera-based system were compared with the ones measured by the POF curvature sensor. We used the POF curvature sensor for the development of the compensation technique, since it was already evaluated with respect to the potentiometer, presenting low errors in this characterization. However, we must emphasize that other sensor systems can be used as reference for the proposed compensation technique, including IMUs, marker-based camera systems and goniometers. The technique proposed here is based on the basic premise that the errors are mainly related to the detection of the parameters d_1, d_2 and d_3 (due to self-occlusions, numerical errors on the computer vision algorithm, among other reasons). Therefore, the errors can be correlated (and then compensated) by considering the actual value of the anthropometric measurements (d_1 and d_2) used on the angle estimation, which can be measured on each subject or estimated using the subject's height [41].

For the first characterization of the technique, the flexion/extension cycles of five subjects (M1, M3, M5, F2 and F5) are analyzed, and a polynomial regression between the angles estimated by the camera-based system and the POF curvature sensor is performed for each of the five subjects, where each equation has the type shown in Equation (7):

$$ang_{ref} = a \cdot ang_{cam}^3 + b \cdot ang_{cam}^2 + c \cdot ang_{cam} + d, \qquad (7)$$

where ang_{cam} is the angle estimated by the camera, ang_{ref} is the angle measured by the POF curvature sensor. In addition, a, b, c and d are polynomial regression coefficients experimentally obtained through the regression between the angular responses of both sensor systems, i.e., markerless camera-based and POF sensor, using the least squares method. The coefficient d is the offset on the sensor response (in °). Therefore, if the sensors responses are normalized in the beginning of the test, the offset will be null. For this reason, the coefficient d is not employed on the analysis of correlation between the coefficients of the angular error correction and the anthropometric measurements of the subjects.

Figure 12 shows the regression between the angle measured by the camera-based system and the POF curvature sensor for the third flexion cycle (as an example) of subject F5. The results show a high correlation (0.998) between the responses using a third-order polynomial regression. Actually, such high correlation occurs for all the cycles of the five subjects analyzed, where the correlation coefficient was higher than 0.9 in all cases. Hence, the assumption of correlation between the errors of both sensor systems holds true (based on the analyses performed). Then, the next step is to correlate the polynomial coefficients (a, b and c) with the anthropometric measurements of each participant.

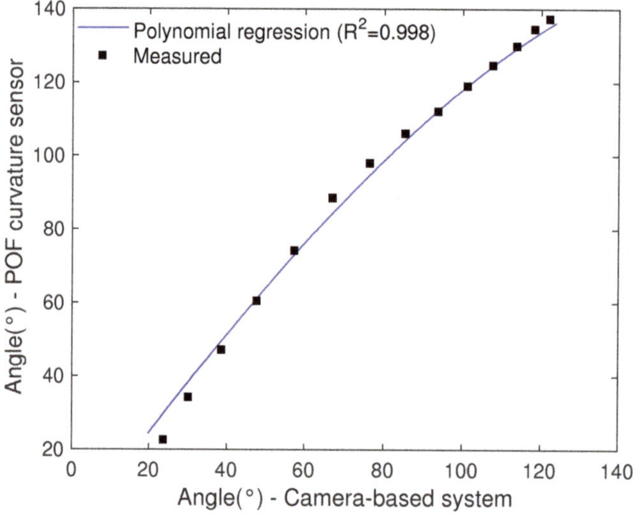

Figure 12. Polynomial regression between camera-based system and POF curvature sensor angular responses for the third flexion cycle of subject F5.

As discussed in Section 2.1, the parameters used on the angle estimation by the camera-based system are the anthropometric distances (d_1 and d_2), which are detected through computer vision algorithms. Thus, errors on the detection of such points will lead to errors on angle estimation, where such errors can be related to those anthropometric distances. However, these parameters (d_1 and d_2) are intrinsic of each subject and can be easily measured. In addition, it is possible to use the height of each subject (as showed in Table 1) in conjunction with anthropometric data for males and females to correlate the arm length with the subject's height. Figure 13 shows the correlation of the polynomial regression coefficients (a, b and c) with the subjects' arm lengths (D).

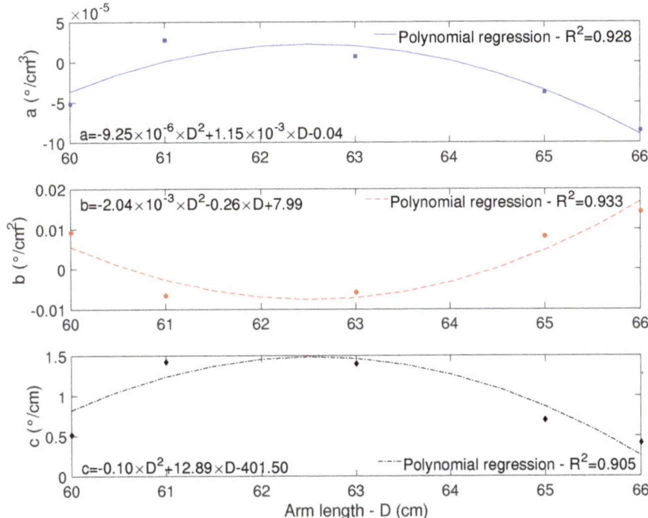

Figure 13. Polynomial regression between the subjects' arm length (parameter D) and the coefficients of angular error correction.

The results as well as the equations presented in Figure 13 indicate the feasibility of using the anthropometric measurements of each subject on equations for angular errors corrections in camera-based systems. The correlation coefficient is higher than 0.9 for all analyzed coefficients, indicating the possibility of using the proposed compensation technique for angle correction. Then, by substituting the equations shown in Figure 13 in Equation (7), it is possible to obtain a corrected angle as depicted in Figure 14 for three flexion/extension cycles for subject F1. In addition, the uncompensated response, i.e., the response of the camera-based system without applying the equations for angle correction, is also presented for comparison purposes. The RMSE for the compensated response is also presented in order to verify the accuracy enhancement provided by the proposed technique. Compared to the uncompensated responses, where the RMSE was 15.04°, 9.25° and 10.23° for cycles 1, 2, and 3, respectively, the proposed angular error compensation was able to reduce the errors substantially in all three cycles. To further verify the performance of the proposed technique, the aforementioned compensation equations were applied for the responses in the sagittal plane for all subjects. The comparison between the RMSEs for the cases with and without the compensation technique is presented in Table 5 for each subject in all three flexion/extension cycles analyzed, where the mean and standard deviation of the three cycles are presented for each participant.

The results presented in Table 5 show the feasibility of the proposed technique, where the RMSE was reduced for all 11 subjects analyzed. The highest reduction occurred in Subject F1, in which the RMSE reduced from 11.52° to 3.52° after applying the correction equations. The mean of the RMSEs for the compensated responses is about 4.90°, whereas the uncompensated one is 10.42°, which means a two-fold reduction of the RMSE when the proposed compensation is applied. It is also worth to mention that the lowest RMSE reduction for the compensated case occurred in subject M3, where the RMSE reduced 2.11°. However, one should note that the RMSE of the uncompensated response of this subject was already low

(6.90°) when compared to the ones of the other subjects and even when compared to the errors presented in the literature for similar systems [40].

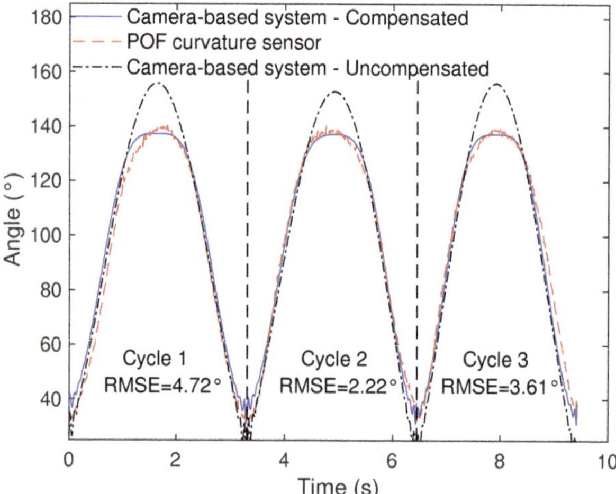

Figure 14. Polynomial regression between the subjects' arm length (parameter d_3) and the coefficients of angular error correction.

Table 5. Comparison between RMSEs of the compensated and uncompensated responses for each subject.

Subject	RMSE-Compensated	RMSE-Uncompensated
M1	3.52° ± 0.84°	10.01° ± 2.12°
M2	3.99° ± 1.15°	10.90° ± 1.67°
M3	4.79° ± 0.74°	6.90° ± 2.25°
M4	6.24° ± 1.66°	12.00° ± 2.83°
M5	4.64° ± 1.23°	10.87° ± 0.15°
F1	3.52° ± 1.25°	11.52° ± 3.24°
F2	4.22° ± 2.25°	10.35° ± 0.60°
F3	4.11° ± 1.31°	7.72° ± 1.59°
F4	5.36° ± 2.05°	12.76° ± 1.63°
F5	5.74° ± 0.77°	8.11° ± 0.81°
F6	4.79° ± 1.06°	13.52° ± 0.95°

The proposed technique for angular errors correction in markerless camera-based system is a feasible and straightforward option to enhance the angular accuracy in such systems. There is a calibration step in which the response of the camera-based system has to be compared with the one of a reference sensor system, e.g., wearable or marker-based camera systems. Then, the errors obtained on the markerless camera-based system are compared with the subject's anthropomorphic parameters (arm length in this case) in order to obtain an equation that relates the angle correction with the parameters of each subject. Therefore, an important caveat should be mentioned: the calibration routine must be performed with respect to a reliable reference, and the movements should be performed at one plane, i.e., sagittal, frontal, or transverse planes movements. In addition, the calibration has to be performed on the same range at which the angle analysis will be performed, i.e., if an angular interval of 0 to 160° will be analyzed, the calibration has to be made at this same angular range (0 to 160°). By following these steps, it is possible to obtain accurate single plane angle measurements with a markerless camera-based system. Therefore, the main limitation of this approach is the necessity of a calibration stage prior to the application of the proposed sensor system in the same range and planes of movement envisaged on the proposed application. However, it is worth noting that the proposed approach can be extended for movement analysis of different degrees of freedom by adjusting the calibration stage accordingly and correlating the errors with the anthropometric parameters of each subject.

4. Conclusions

This paper presented the analysis and comparison of a markerless camera-based system for elbow angles assessment. The proposed markerless system uses two RGB-D cameras to reduce errors and inaccuracies related to self-occlusion issues. The non-wearable system performance was compared with two wearable solutions, namely POF curvature sensor and IMUs, in flexion/extension movements performed in different planes (sagittal, transverse, and frontal planes). Even though the proposed markerless camera-based system showed lower errors than some similar systems proposed in the literature, the errors are still high for movement analysis applications. To tackle this limitation, a comprehensive analysis of the system showed that despite the high errors, the markerless camera-based system response has a high correlation with the response of the wearable sensors. This also indicated the possibility of applying post-processing techniques aimed at error reduction in those systems. Thus, a compensation technique based on anthropometric measurements of the subjects was proposed and validated using the POF curvature sensor measures, resulting in a significant decrease of the RMSE.

The proposed RGB-D vision system and the novel compensation technique proposed here indicate the suitability of the system on the movement analysis, since the mean error obtained is about 4° for the angles tested, i.e., up to 120° in the sagittal plane, which is in agreement with the errors obtained in some sensing approaches for movement analysis [42]. Thus, this work can pave the way for movement analysis applications with markerless camera-based system. Future works include the further investigation of this technique in the other motion planes (frontal and transverse plane) and also in 3D movement scenarios. In addition, further evaluation and new sensor fusion techniques of markerless camera-based systems and optical fiber sensors will also be developed. Currently, to decrease the error, this work can be understood as the foundation of such developments.

Author Contributions: N.V.-J., L.V.-V., P.C.-R. implemented the assessment protocol and perform all the tests. N.V.-J., L.A., A.A.R.-D., M.L. implemented the RGB-D system proposed, analyzed the data, contributed in paper writing and revisions. A.L.-J. implemented the proposed POF curvature sensor system, analyzed the data, conceived and implemented the angle correction technique, contributed in paper writing and revisions. L.V.-V., P.C.-R. implemented the IMUs system proposed, contributed in paper writing and revisions. C.M., A.F., T.B. assisted in careful reviewing of the paper and proposed various refinements to the draft proposal made.

Funding: This research is financed by CAPES (88887.095626/2015-01)—financing code 001, FAPES (72982608), CNPq (304192/2016-3) and Innovaccion Cauca Research Project-02-2014 Doctorados Nacionales. This research is also financed by FCT through the program UID/EEA/50008/2019 by the National Funds through the Fundação para a Ciência e a Tecnologia / Ministério da Educação e Ciência, and the European Regional Development Fund under the PT2020 Partnership Agreement. This work is also funded by national funds (OE), through FCT – Fundação para a Ciência e a Tecnologia, I.P., in the scope of the framework contract foreseen in the numbers 4, 5 and 6 of the article 23, of the Decree-Law 57/2016, of August 29, changed by Law 57/2017, of July 19

Conflicts of Interest: The authors declare no conflict of interest.

Abbreviations

The following abbreviations are used in this manuscript:

RGB-D   Red-Blue-Green-and-Depth information
IMU   Inertial Measurement Unit
POF   Polymer Optical Fiber
DOF   Degree of Freedom
MEM   Micro-Electro-Mechanical
ROS   Robot Operating System
XML   eXtensible Markup Language
RPC   Remote Procedure Call
NTP   Network Time Protocol

References

1. Hashimoto, K.; Higuchi, K.; Nakayama, Y.; Abo, M. Ability for basic movement as an early predictor of functioning related to activities of daily living in stroke patients. *Neurorehabilit. Neural Repair* **2007**, *21*, 353–357. [CrossRef] [PubMed]
2. Casamassima, F.; Ferrari, A.; Milosevic, B.; Ginis, P.; Farella, E.; Rocchi, L. A wearable system for gait training in subjects with Parkinson's disease. *Sensors* **2014**, *14*, 6229–6246. [CrossRef] [PubMed]
3. Van Den Noort, J.C.; Ferrari, A.; Cutti, A.G.; Becher, J.G.; Harlaar, J. Gait analysis in children with cerebral palsy via inertial and magnetic sensors. *Med. Biol. Eng. Comput.* **2013**, *51*, 377–386. [CrossRef] [PubMed]
4. Muro-de-la Herran, A.; García-Zapirain, B.; Méndez-Zorrilla, A. Gait analysis methods: An overview of wearable and non-wearable systems, highlighting clinical applications. *Sensors* **2014**, *14*, 3362–3394. [CrossRef] [PubMed]
5. Shull, P.B.; Jirattigalachote, W.; Hunt, M.A.; Cutkosky, M.R.; Delp, S.L. Quantified self and human movement: A review on the clinical impact of wearable sensing and feedback for gait analysis and intervention. *Gait Posture* **2014**, *40*, 11–19. [CrossRef] [PubMed]
6. Wong, C.; Zhang, Z.Q.; Lo, B.; Yang, G.Z. Wearable Sensing for Solid Biomechanics: A Review. *IEEE Sens. J.* **2015**, *15*, 2747–2760. [CrossRef]
7. Hawkins, D. A new instrumentation system for training rowers. *J. Biomech.* **2000**, *33*, 241–245. [CrossRef]
8. Wang, P.T.; King, C.E.; Do, A.H.; Nenadic, Z. A durable, low-cost electrogoniometer for dynamic measurement of joint trajectories. *Med. Eng. Phys.* **2011**, *33*, 546–552. [CrossRef]
9. El-Gohary, M.; McNames, J. Shoulder and Elbow Joint Angle Tracking with Inertial Sensors. *IEEE Trans. Biomed. Eng.* **2012**, *59*, 2635–2641. [CrossRef]
10. Peters, K. Polymer optical fiber sensors—A review. *Smart Mater. Struct.* **2011**, *20*, 013002. [CrossRef]
11. Leal-Junior, A.G.; Frizera, A.; Vargas-Valencia, L.; Dos Santos, W.M.; Bo, A.P.; Siqueira, A.A.; Pontes, M.J. Polymer Optical Fiber Sensors in Wearable Devices: Toward Novel Instrumentation Approaches for Gait Assistance Devices. *IEEE Sens. J.* **2018**, *18*, 7085–7092. [CrossRef]
12. Leal-Junior, A.; Theodosiou, A.; Díaz, C.; Marques, C.; Pontes, M.J.; Kalli, K.; Frizera-Neto, A. Polymer optical fiber Bragg Gratings in CYTOP Fibers for angle measurement with dynamic compensation. *Polymers* **2018**, *10*, 674. [CrossRef]
13. Gong, Y.; Zhao, T.; Rao, Y.J.; Wu, Y. All-fiber curvature sensor based on multimode interference. *IEEE Photonics Technol. Lett.* **2011**, *23*, 679–681. [CrossRef]
14. Bilro, L.; Alberto, N.; Pinto, J.L.; Nogueira, R. Optical sensors based on plastic fibers. *Sensors* **2012**, *12*, 12184–12207. [CrossRef] [PubMed]
15. Leal-Junior, A.; Frizera, A.; Marques, C.; José Pontes, M. Polymer-optical-fiber-based sensor system for simultaneous measurement of angle and temperature. *Appl. Opt.* **2018**, *57*, 1717–1723. [CrossRef] [PubMed]

16. Wu, G.; Van Der Helm, F.C.; Veeger, H.E.; Makhsous, M.; Van Roy, P.; Anglin, C.; Nagels, J.; Karduna, A.R.; McQuade, K.; Wang, X.; et al. ISB recommendation on definitions of joint coordinate systems of various joints for the reporting of human joint motion—Part II: Shoulder, elbow, wrist and hand. *J. Biomech.* **2005**, *38*, 981–992. [CrossRef]

17. Menezes, R.C.; Batista, P.K.; Ramos, A.Q.; Medeiros, A.F. Development of a complete game based system for physical therapy with kinect. In Proceedings of the 2014 IEEE 3nd International Conference on Serious Games and Applications for Health (SeGAH), Rio de Janeiro, Brazil, 14–16 May 2014; pp. 1–6.

18. Svendsen, J.; Albu, A.B.; Virji-Babul, N. Analysis of patterns of motor behavior in gamers with down syndrome. In Proceedings of the IEEE Computer Society Conference on Computer Vision and Pattern Recognition Workshops, Colorado Springs, CO, USA, 20–25 June 2011; doi:10.1109/CVPRW.2011.5981673.

19. Cameirao, M.S.; Bermudez i Badia, S.; Duarte Oller, E.; Verschure, P.F. Neurorehabilitation using the virtual reality based Rehabilitation Gaming System: Methodology, design, psychometrics, usability and validation. *J. Neuroeng. Rehabil.* **2010**, *7*, 48. [CrossRef]

20. Müller, B.; Ilg, W.; Giese, M.A.; Ludolph, N. Validation of enhanced Kinect sensor based motion capturing system for gait assessment. *PLoS ONE* **2017**, *12*, e0175813. [CrossRef]

21. Konstantinidis, E.I.; Billis, A.S.; Paraskevopoulos, I.T.; Bamidis, P.D. The interplay between IoT and serious games towards personalised healthcare. In Proceedings of the 9th International Conference on Virtual Worlds and Games for Serious Applications (VS-Games), Athens, Greece, 6–8 September 2017; pp. 249–252. [CrossRef]

22. Covaci, A.; Kramer, D.; Augusto, J.C.; Rus, S.; Braun, A. Assessing Real World Imagery in Virtual Environments for People with Cognitive Disabilities. In Proceedings of the 2015 International Conference on Intelligent Environments, Prague, Czech Republic, 15–17 July 2015; pp. 41–48.

23. Abellard, A.; Abellard, P.; Applications, A.H. Serious Games Adapted to Children with Profound Intellectual and Multiple Disabilities. In Proceedings of the 9th International Conference on Virtual Worlds and Games for Serious Applications (VS-Games), Athens, Greece, 6–8 September 2017; pp. 183–184.

24. Brandão, A.; Trevisan, D.G.; Brandão, L.; Moreira, B.; Nascimento, G.; Vasconcelos, C.N.; Clua, E.; Mourão, P.T. Semiotic inspection of a game for children with Down syndrome. In Proceedings of the 2010 Brazilian Symposium on Games and Digital Entertainment, Florianopolis, Brazil, 8–10 November 2010; pp. 199–210. [CrossRef]

25. Chen, P.J.; Du, Y.C.; Shih, C.B.; Yang, L.C.; Lin, H.T.; Fan, S.C. Development of an upper limb rehabilitation system using inertial movement units and kinect device. In Proceedings of the 2016 International Conference on Advanced Materials for Science and Engineering (ICAMSE), Tainan, Taiwan, 12–13 November 2016; pp. 275–278.

26. Lee, J.H.; Nguyen, V.V. Full-body imitation of human motions with kinect and heterogeneous kinematic structure of humanoid robot. In Proceedings of the 2012 IEEE/SICE International Symposium on System Integration (SII), Fukuoka, Japan, 16–18 December 2012; pp. 93–98.

27. Qi, Y.; Soh, C.B.; Gunawan, E.; Low, K.S.; Maskooki, A. A novel approach to joint flexion/extension angles measurement based on wearable UWB radios. *IEEE J. Biomed. Health Inform.* **2014**, *18*, 300–308.

28. Xu, Y.; Yang, C.; Zhong, J.; Wang, N.; Zhao, L. Robot teaching by teleoperation based on visual interaction and extreme learning machine. *Neurocomputing* **2018**, *275*, 2093–2103. [CrossRef]

29. Leal Junior, A.G.; Frizera, A.; Pontes, M.J. Analytical model for a polymer optical fiber under dynamic bending. *Opt. Laser Technol.* **2017**, *93*, 92–98. [CrossRef]

30. Zubia, J.; Arrue, J. Plastic optical fibers: An introduction to their technological processes and applications. *Opt. Fiber Technol.* **2001**, *7*, 101–140. [CrossRef]

31. Leal-Junior, A.G.; Frizera, A.; José Pontes, M. Sensitive zone parameters and curvature radius evaluation for polymer optical fiber curvature sensors. *Opt. Laser Technol.* **2018**, *100*, 272–281. [CrossRef]

32. Paulich, M.; Schepers, H.M.; Rudigkeit, N.; Bellusci, G. *Xsens MTw Awinda: Miniature Wireless Inertial-Magnetic Motion Tracker for Highly Accurate 3D Kinematic Applications*; Technical Report; XSens Technologies: Enschede, The Netherlands, 2018.

33. Vargas-Valencia, L.S.; Elias, A.; Rocon, E.; Bastos-Filho, T.; Frizera, A. An IMU-to-Body Alignment Method Applied to Human Gait Analysis. *Sensors* **2016**, *16*, 2090. [CrossRef] [PubMed]

34. Cutti, A.G.; Giovanardi, A.; Rocchi, L.; Davalli, A.; Sacchetti, R. Ambulatory measurement of shoulder and elbow kinematics through inertial and magnetic sensors. *Med. Biol. Eng. Comput.* **2008**, *46*, 169–178. [CrossRef] [PubMed]
35. Palermo, E.; Rossi, S.; Marini, F.; Patanè, F.; Cappa, P. Experimental evaluation of accuracy and repeatability of a novel body-to-sensor calibration procedure for inertial sensor-based gait analysis. *Measurement* **2014**, *52*, 145–155. [CrossRef]
36. Laidig, D.; Müller, P.; Seel, T. Automatic anatomical calibration for IMU-based elbow angle measurement in disturbed magnetic fields. *Curr. Dir. Biomed. Eng.* **2017**, *3*, 167–170. [CrossRef]
37. Seel, T.; Raisch, J.; Schauer, T. IMU-Based Joint Angle Measurement for Gait Analysis. *Sensors* **2014**, *14*, 6891–6909. [CrossRef]
38. Tannous, H.; Istrate, D.; Benlarbi-Delai, A.; Sarrazin, J.; Gamet, D.; Ho Ba Tho, M.C.; Dao, T.T. A new multi-sensor fusion scheme to improve the accuracy of knee flexion kinematics for functional rehabilitation movements. *Sensors* **2016**, *11*, 1914. [CrossRef]
39. Kirtly, C. *Clinical Gait Analysis: Theory and Practice*; Churchill Livingstone: London, UK, 2006; Volume 3, p. 2007.
40. Schmitz, A.; Ye, M.; Shapiro, R.; Yang, R.; Noehren, B. Accuracy and repeatability of joint angles measured using a single camera markerless motion capture system. *J. Biomech.* **2014**, *47*, 587–591. [CrossRef]
41. Gordon, C. Anthropometric Detailed Data Tables. 2006. Available online: https://multisite.eos.ncsu.edu/www-ergocenter-ncsu-edu/wp-content/uploads/sites/18/2016/06/Anthropometric-Detailed-Data-Tables.pdf (accessed on 26 November 2018).
42. Piriyaprasarth, P.; Morris, M.E. Psychometric properties of measurement tools for quantifying knee joint position and movement: A systematic review. *Knee* **2007**, *14*, 2–8. [CrossRef] [PubMed]

Article

Neural Spike Digital Detector on FPGA

Elia Arturo Vallicelli [1,*], Marco Reato [2], Marta Maschietto [2], Stefano Vassanelli [2], Daniele Guarrera [3], Federico Rocchi [2], Gianmaria Collazuol [3], Ralf Zeitler [4], Andrea Baschirotto [1] and Marcello De Matteis [1,*]

1   Department of Physics, University of Milano Bicocca, 20126 Milano, Italy; andrea.baschirotto@unimib.it
2   Department of Biomedical Sciences, University of Padova, 35131 Padova, Italy;
    marco.reato.88@gmail.com (M.R.); marta.maschietto@unipd.it (M.M.); stefano.vassanelli@unipd.it (S.V.);
    rocchi3d@yahoo.com (F.R.)
3   Department of Physics, University of Padova, 35131 Padova, Italy;
    daniele.guarrera@studenti.unipd.it (D.G.); gianmaria.collazuol@unipd.it (G.C.)
4   Venneos GmbH, 70569 Stuttgart, Germany; zeitler@venneos.com
*   Correspondence: e.vallicelli@campus.unimib.it (E.A.V.); marcello.dematteis@unimib.it (M.D.M.)

Received: 31 October 2018; Accepted: 30 November 2018; Published: 5 December 2018

Abstract: This paper presents a multidisciplinary experiment where a population of neurons, dissociated from rat hippocampi, has been cultivated over a CMOS-based micro-electrode array (MEA) and its electrical activity has been detected and mapped by an advanced spike-sorting algorithm implemented on FPGA. MEAs are characterized by low signal-to-noise ratios caused by both the contactless sensing of weak extracellular voltages and the high noise power coming from cells and analog electronics signal processing. This low SNR forces to utilize advanced noise rejection algorithms to separate relevant neural activity from noise, which are usually implemented via software/off-line. However, off-line detection of neural spikes cannot be obviously used for real-time electrical stimulation. In this scenario, this paper presents a proper FPGA-based system capable to detect in real-time neural spikes from background noise. The output signals of the proposed system provide real-time spatial and temporal information about the culture electrical activity and the noise power distribution with a minimum latency of 165 ns. The output bit-stream can be further utilized to detect synchronous activity within the neural network.

Keywords: biological neural networks; biosensors; neural engineering; digital circuits; field programmable gate arrays; principal component analysis

---

1. Introduction

The communication between neurons is carried out through action potentials (AP), transient changes of a trans-membrane voltage of about 100 mV$_{PP}$ and few kHz bandwidth. High spatial resolution detection of AP signals (thousands of recording sites/pixels) opens an unthinkable scenario in neuroscience since it allows to observe simultaneously large populations of neurons and their communications [1–4]. For several years, the most common technique to observe neurons utilized needle-shaped probes that deeply penetrate the cells, reducing the average-life of the cells to few days due to the irreversible damage caused to the pierced neural membrane but can extract AP signals at high signal-to-noise ratio (SNR [5–7]). Recently, state-of-the-art approaches have been widely exploiting minimally-invasive sensing techniques, which are based on CMOS microelectrode arrays (MEAs [8–10]). This avoids penetration of the neural membrane and therefore greatly limits tissue damage, allowing months-long observations of neural cultures and opening the road to long-term implants [1]. However, SNR is heavily reduced due to the lower signal power of extracellular signals.

For this reason, recorded signals are typically improved by advanced post-processing spike sorting algorithms that separate relevant neural spikes (deterministic events) from background noise (random fluctuations), even in presence of very low SNR [9,11]. Thus, they deeply analyze the behavior of neurons populations [11], whereas a complete digital processing hardware implementation is needed to apply event-driven electrical stimulation techniques (e.g., deep brain stimulation) [12].

Therefore, closed-loop neurons stimulation intrinsically needs advanced electronics systems composed by:

- analog stages to acquire the biological signals and perform the conversion into digital domain;
- advanced digital spike-sorting algorithms to separate in real-time AP signals from background noise;
- electrical stimulation stages that interact with neurons in response to their activity

This work analyzes the digital spike-sorting hardware design. More specifically, a complete action potential detector has been implemented on a Xilinx Spartan 6 FPGA (XC6SLX45-2C, -2 speed grade) [13] that includes a noise-rejection algorithm based on principal component analysis (PCA [9]). The hereby proposed neural spike digital detector (NSDD) identifies single AP signals with amplitudes around 200–600 µV that are recorded with a 16 × 192 pixel-submatrix from a 256 × 384 pixel CMOS MEA with an average noise power per pixel of about 100 μV_{RMS}. It maps in real-time the neural culture electrical activity in terms of total AP number of events (over a time width of 15 s acquisition), AP frequency (enabling the detection of synchronous pattern), and noise power spatial distribution.

Hardware implementation of spike sorting algorithms is very important for implanted devices because of their intrinsic limited power budget. Performing neural activity recognition in situ greatly limits the bandwidth required to transmit data from the MEA and therefore the associated power consumption, at the cost of some additional digital circuitry which however consumes very little power in scaled technologies.

This paper is organized as follows. Section 2 introduces the experimental setup composed by the neuronal cells culture, the electronics read-out for neural potential sensing/digitalization and the FPGA NSDD. Section 3 illustrates how the digital spike sorting algorithm has been implemented on FPGA. Finally, Section 4 provides experimental results in terms of neural network electrical activity mapping, noise spatial distribution and spatial detection of synchronous neural activity. At the end of the paper, conclusions will be drawn.

2. Neural Spike Digital Detector (NSDD)

Figure 1 illustrates the top-level block-scheme of the experimental setup described in this work. The neuronal cells dissociated from rat hippocampi are seeded on the CMOS MEA composed by two main stages: the sensor matrix (based on electrolyte-oxide–metal-oxide-semiconductor (EOMOS) transistors) that capacitively senses the neuronal cells electrical activity and the electronics signal processing stages (analog front-end in Figure 1).

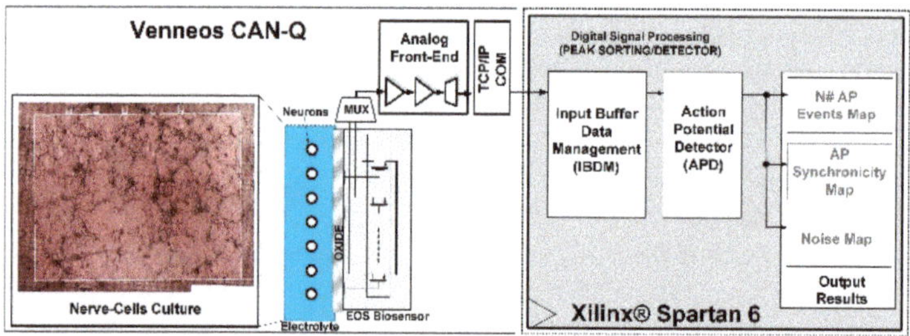

Figure 1. Neuronal-cells culture experimental setup [3].

The CMOS MEA is organized in a matrix of 256 × 384 pixels (for an overall 98,304 pixels spatial resolution). The pixels are organized in a hexagonal grid with 6.5 × 5.8 µm pitch. 16 × 192 pixels are time-division-multiplexed to one of 32 analog output channels. The 32 output signals are off-chip digitized.

A communication interface (TCP/IP COM) forwards the digitized signals from one selected output to the NSDD FPGA at 14.1 MS/s. The single pixel signal time evolution is shown in Figure 2 (0.1 s time window and at 4.6 kSample/s).

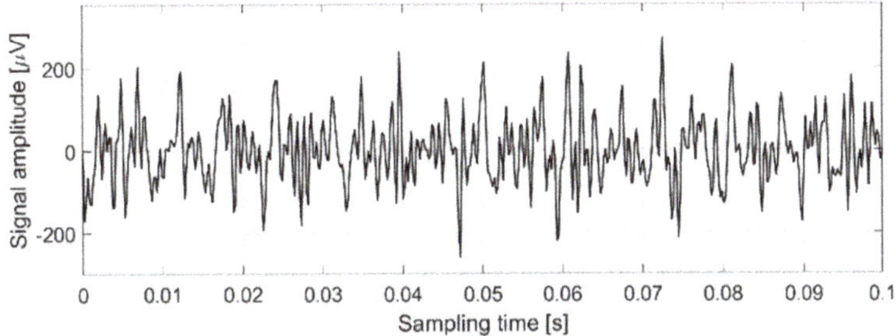

Figure 2. Single pixel signal time evolution.

2.1. Neuronal Cell Culture

A top-view microscope image of the neuronal cells culture is shown in Figure 3. After dissociation from the tissue, the hippocampal neurons in culture are able to reconstruct a vital network with specific connections, thus being useful in validating the performance of devices for electrophysiological signal detection.

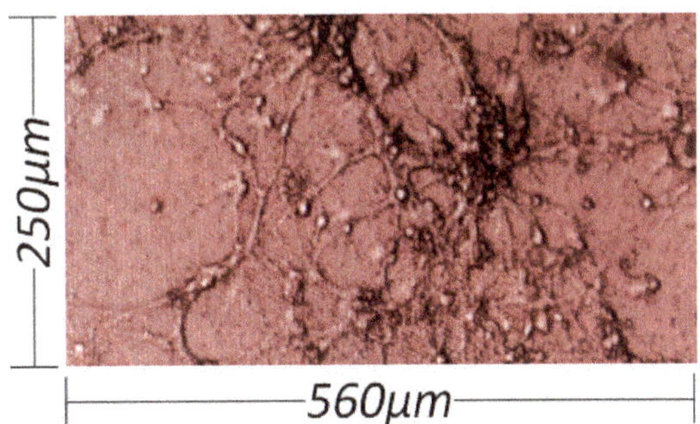

Figure 3. Example of hippocampal neurons culture on the chip.

Wistar rats (Charles River) are maintained in the Animal Research Facility of the Department of Biomedical Sciences—University of Padova. All the procedures involving animals are realized according to Italian and local regulation concerning animal welfare (OPBA Unipd and dL 26/2014). All the reagents and media are from Gibco (Thermo Fisher Scientific - Life Technologies Italia, Monza, Italy). Neuronal cells are dissociated from the hippocampi of E18–E19 embryos, as previously described [10]. Briefly, the dissected hippocampi are digested in 0.125% Trypsin for 20 min at 37 °C and then dissociated to a single cell suspension in complete DMEM-GlutaMAX-1 (medium supplemented with 10% FBS, 1 u/mL penicillin and 1 μg/mL streptomycin). After centrifugation at 250 g for 10 min, the pellet is gently re-suspended in complete DMEM and pre-plated onto a cell culture dish for 2 h in an incubator at 37 °C and 5% (v/v) CO_2 to reduce the percentage of glial cells in the final culture. The collected supernatant is then centrifuged at 250 g for 10 min and the pellet re-suspended in complete DMEM. The surface of the sensors is coated with a proteic layer of poly-L-lysine, that helps neuronal adhesion and differentiation. About 150,000 neurons/cm^2 are seeded on the chip in L15 medium added with 5% FBS and 1% penicillin/streptomycin for 2 h and then maintained in the incubator in complete NeuroBasal medium (added with 2% B27 supplement and 1% GlutaMAX-1). Pyramidal (excitatory) neurons are usually more represented in culture than other type of neurons. The mature neuronal network presents an extensive branching with large amount of synaptic connections [9]. The experiments are realized on mature cultures at DIV (days in vitro) 20–30.

2.2. Neural Interface Noise

The CMOS MEAs sense the extracellular voltage induced by neurons transmembrane ionic currents. This saves biological tissues from disruptive penetration, does not prevent biological regeneration and finally enable long-time cell observation. Unfortunately, extracellular signal amplitudes are two orders of magnitude smaller compared to more invasive techniques (i.e., intracellular needles probes) and therefore SNR is lower. Moreover, adhesion of cells to the surface increases the noise power [6]. For this reason, accurate sensing of neural cells electrical activity needs an accurate spatial evaluation of the noise power.

The noise power of the signal has been obtained by analyzing the experimental data coming from a 2.5 s time window of the culture activity, obtaining the noise power spatial map vs. pixels in Figure 4 (not in scale). It is possible to identify two principal areas, depending on the noise power level: areas without cells exhibit about 80 μV_{RMS} noise power (blue regions in Figure 4), while the pixels beneath the cells have higher noise power (120 μV_{RMS}, yellow/red in Figure 4).

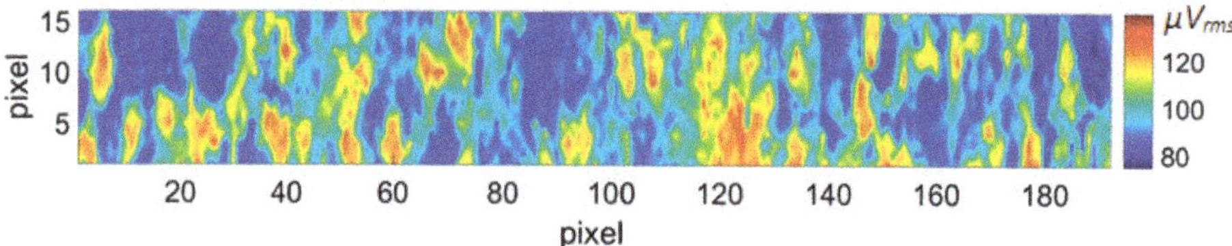

Figure 4. Noise power spatial map (over 2.5 s acquisition time, noise is integrated over 4 kHz bandwidth) [3].

The difficulty to separate the AP from noise is also evident in Figure 2 that shows the single pixel time evolution.

Due to the high noise power, using a simple 5·σ-threshold-crossing detector is not a viable option [11], because it would lead to discarding most of the action potentials. In particular, with reference to Figure 2, a 5·σ-detector at 120 μV_{RMS} noise would lead to a 600 μV threshold voltage and hence jump over several relevant action potentials events.

2.3. Principal Component Analysis

MEAs are characterized by high spatial and temporal resolution that can be exploited by correlation algorithms such as principal component analysis (PCA) [11]. The pixel density and the sample frequency are sufficiently high that a single AP deterministic event can be detected by nine adjacent pixels for three consecutive time samples (3 × 3 pixel square, see 9PIX-Sub-Set in Figure 5).

Thus, a single AP can be observed in a total of 27 samples in close spatial/temporal proximity. The PCA algorithm is used to calculate the probability to detect AP events in front of noise (statistical thermal fluctuations).

The AP detection implies to comply with the condition [9,11]

$$\sum_{pixj=1\to9} \frac{\sum_{\{n,n-1,n-2\}} PIXj(n)^2}{\sigma_{NOISE,j}^2} \geq AP\_Threshold \cong 84.6 \tag{1}$$

where PIX_j indicates the j-th pixel (sampled at 3 different instants: n, $n-1$ and $n-2$).

The signal is encoded in a digital word of 14 bits. $\sigma_{NOISE,j}^2$ is the calculated noise power over the 3×3 matrix. Therefore, to separate the eventual AP spike from noise, the algorithm takes nine-pixel sub-set (9PIX-Sub-set as indicated in Figure 5).

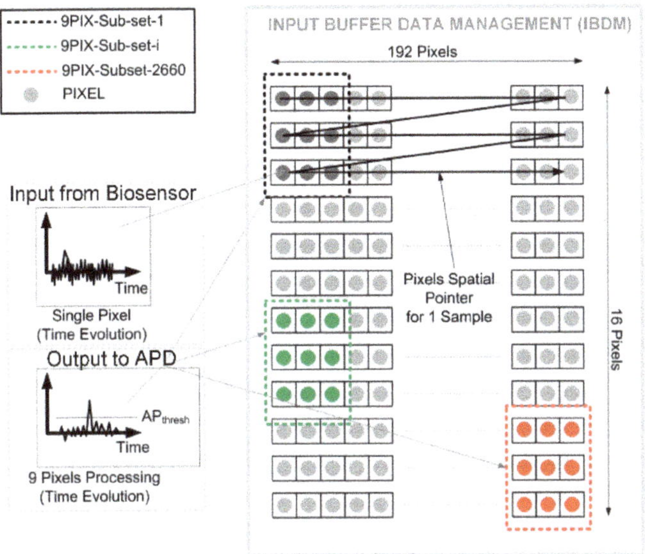

Figure 5. Input buffer data management functional scheme.

The square sum of three consecutive time-samples of nine adjacent pixels is divided by the noise (i.e., $\frac{\sum_{\{n,n-1,n-2\}} PIXj(n)^2}{\sigma_{NOISE,j}^2}$) and is then compared with the AP detection threshold (*AP_Threshold*). The *AP_Threshold* value is calculated imposing that the whole MEA produces one false positive event per second due to noise random fluctuation. Using a MEA subset of 16×192 pixels at 4.6 KS/s sampling rate per pixel, the calculated *AP_Threshold* is 84.6.

Such a technique can improve the global SNR for a sub-set of nine pixels, at the cost of a small spatial resolution reduction (the final spatial map will have 14×190 equivalent extracted pixels, since two generic nine-pixel subsets share two columns and are thus partially overlapped). Figure 6 shows the PCA output for a specific sub-set of nine pixels, compared with the threshold value. After the PCA processing the SNR increases. It goes from 1 dB (without PCA) up to 7 dB (with PCA).

Thus, it is now more reliable to apply a threshold crossing approach (*threshold* value is placed at 84.6) and to detect two AP events at 1.9 s and 2.2 s. On average, when no AP occurs, the signal RMS power for each pixel will be equal to $\sigma_{NOISE,j}$. Therefore, on average $\frac{PIXj(t)^2}{\sigma_{NOISE,j}^2}$ will be equal to 1, and the combined 27-samples value $\sum_{pixj=1 \to 9} \frac{\sum_{t=0,1,2} PIXj(t)^2}{\sigma_{NOISE,j}^2}$ will be 27. The AP detection condition implies that an AP is detected when the PCA output signal (Figure 6) is higher than 84.6. This leads to a minimum SNR value of at least 7 dB (2).

$$SNR_{dB} = 20log_{10}\left(\frac{84.6}{27}\right) - 3dB \cong 7dB \tag{2}$$

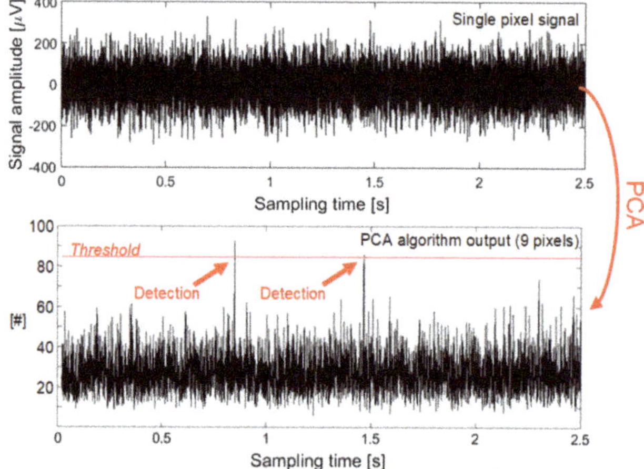

Figure 6. PCA output (left side of Equation (1)) compared with threshold.

3. FPGA Neural Spike Digital Detector

The PCA algorithm has been implemented on FPGA by a dedicated digital design that is interfaced with the neurons culture in order to detect and separate APs spikes from noise. The system implementing the PCA is embedded in the NSDD system in Figure 1 and it also manages the MEA output data rate performing a real time detection of the relevant APs. Thus, it operates by three main stages:

- the input-buffer-data-management (IBDM) synchronously receives the data coming from the biosensor via the TCP/IP communication interface;
- the action potential detector (APD), that is the digital circuit implementing the PCA algorithm;
- a specific set of MATLAB functions that provide graphical representation of the on-going neural activity.

The main requirement for a real time processing of the signal coming from the MEA is that the NSDD data throughput must be equal or higher than the MEA sample rate.

This translates in the Equation (3) condition that correlates the MEA sample rate ($MEA_{OUT,RATE}$), the FPGA master clock frequency (f_{CLK}) and the total number of clock cycles needed to process one sample (N_{CLK})

$$f_{CLK} \geq N_{CLK} \cdot MEA_{OUT,RATE} \tag{3}$$

In this way, the APD throughput capability is able to manage the input data rate. This approach has the following advantages:

- no data pile-up occurs in the input buffer;
- PCA algorithm on-line runs for long periods of time with minimum buffer size requirements.

3.1. Input Buffer Data Management

The NSDD receives a stream of data from the Venneos-CAN-Q machine (VCQ [14]) via an Ethernet connection using TCP/IP protocol. Data are acquired by the VCQ from the MEA matrix using a specific raster scanning algorithm, whose general procedure is shown in Figure 5.

Each pixel of the MEA is sampled at 4.6 kS/s and data from the whole pixel matrix are then time-division-multiplexed into a single data-stream that is sent via TCP/IP COM to the input buffer data management stage. The multiplexed data-stream is obtained by transmitting the first time-sample acquired by each pixel, starting from the top-left pixel (first data) and scanning the whole matrix with a raster pattern, thus one-pixel-by-one from left to right and one-line-by-one from top to bottom. After the whole MEA is scanned for the first time-frame, the raster scan is repeated for the subsequent time frames.

The MEA is composed by 3072 pixels and therefore for the first 3072 clock cycles the NSDD receives the first time-sample of each pixel. During the second 3072 clock cycles the NSDD receives the second time-sample of each pixel and then the TCP/IP COM follows this procedure up to the last sample of the last pixel.

The IBDM manages the input data-stream by controlling the Ethernet interface and storing the received data in a dedicated buffer. Then, it sends a data-stream from the buffer to the APD stage to perform the PCA algorithm.

Effectively, Equation (1) establishes that three consecutive time-samples for nine adjacent pixels (27 samples in total) will be used for thresholding. Therefore, the APD can compute the PCA on the first nine-pixel sub-set (corresponding to 9PIX-Sub-set1 in Figure 5) only after the MEA has been scanned completely two times (3072 + 3072 clock cycles) and the VCQ sends the third time-sample of each pixel of the first two rows (192 + 192 clock cycles) plus the third time-sample of the first three pixels of the third row. Therefore, the first output data comes after an initial start-up time of 447.3 μs (corresponding to a total of 6531 clock periods) that are needed to sense and transfer to the APD all the 27 samples from 9PIX-Sub-set1. After this start-up time, the PCA is performed on one 9PIX-Sub-set per clock cycle, and the 9PIX-Sub-set is moved on the MEA matrix in a raster scan pattern, synchronous to the input data-stream. Thus, after the start-up time latency, the system on FPGA will operate in real-time providing spike detection at the same TCP/IP sample rate (14.1 MHz).

3.2. Action Potential Detector Algorithm

The action potential detector is that stage in the system implementing the PCA algorithm. A flowchart of the FPGA-APD operations is shown in Figure 7.

Each time a new data ($PIX_j(n)$, is received from the MEA, the APD executes the operations illustrated in the flowchart. The algorithm starts reading the input data up to providing an output data. Output data encodes whether an AP has been detected in the n-th timeframe (time pointer) and in the j-th pixel (spatial pointer).

The APD performs all the operations that are required to verify the specific condition expressed in Equation (1). First it calculates the square sum of three consecutive time-samples from the same pixel

$$OUT\_\Sigma1 = \sum_{\{n,n-1,n-2\}} PIXj(n)^2 \tag{4}$$

$PIX_j(n)$, $PIX_j(n-1)$, $PIX_j(n-2)$ are 16-bits data words coming from IBDM. They encode the n-th time-samples of the j-th pixel. Hence j is the pixel spatial pointer ($j = 1, \dots , 3072$).

Secondly, it calculates the OUT_DIV number, which is the result of the division between the $OUT\_\Sigma1$ and the RMS noise ($\sigma^2_{NOISE}(j)$) of the j-th pixel

$$OUT\_DIV = \frac{OUT\_\Sigma1}{\sigma_{NOISE,j}^2} = \frac{\sum_{\{n,n-1,n-2\}} PIXj(n)^2}{\sigma_{NOISE,j}^2} \tag{5}$$

$\sigma^2_{NOISE}(j)$ is offline calculated for every pixel, and stored in the σ^2-Register.

The OUT_DIV data is then stored into an intermediate buffer (IMB) until all OUT_DIV values of a nine-pixel sub-set are available. Then the sum of OUT_DIV over nine adjacent pixels is calculated

$$OUT\_\Sigma2 = \sum_{j=1\rightarrow9} OUT\_DIVj = \sum_{j=1\rightarrow9} \frac{\sum_{\{n,n-1,n-2\}} PIXj(n)^2}{\sigma_{NOISE,j}^2} \tag{6}$$

The *OUT_Σ2* value is compared with *AP_Threshold* (stored in a specific 16 bits register) to determine whether an AP has occurred or not

$$AP\_Bit = \begin{cases} '1'\ if\ \sum_{pixj=1\rightarrow9} \dfrac{\sum_{\{n,n-1,n-2\}} PIXj(n)^2}{\sigma_{NOISE,j}^2} \geq AP\_Threshold \\ '0'\ \ \ \ otherwise \end{cases} \tag{7}$$

AP_Bit is at '1' logic level if Equation (7) is true, meaning that an action potential has been detected on the 3 × 3 pixel sub-set in the three consecutive timeframes. Otherwise, it is at '0' logic level and no neural activity has been detected.

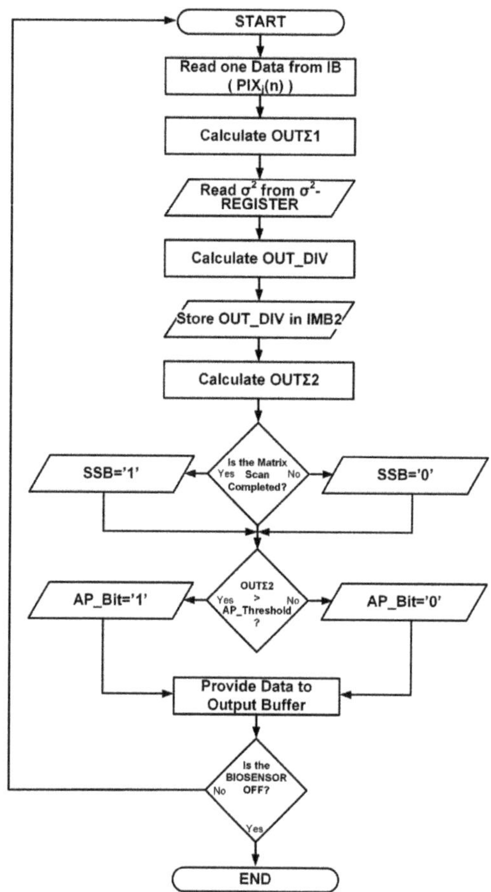

Figure 7. Flowchart of implemented PCA algorithm.

Notice that the algorithm uses a spatial bit/flag (called spatial synchronicity bit (SSB)) whose time evolution w.r.t. the MEA input data stream is qualitatively shown in Figure 8. The SSB flag is high when the input sample comes from the first pixel of the matrix, otherwise it is low.

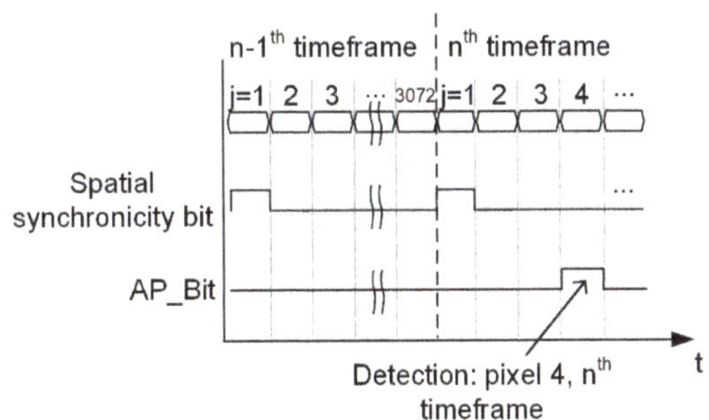

Figure 8. Output signal encoded spatial and temporal information.

Figure 9 shows the AP_Bit and the SSB time evolution extracted from the FPGA output pins using Agilent 16821A logic analyzer [15]. The input pattern signals have been directly taken from the experimental setup in Figure 1.

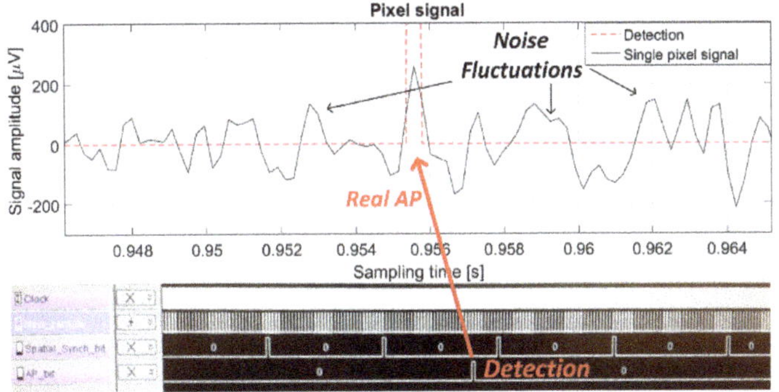

Figure 9. AP detection and single pixel signal with AP.

From top to bottom, the signals recorded by the logic analyzer are:

- the FPGA master clock;
- the IBDM_ENABLE signal which is at '1' logic level when a new input sample is received from the IBDM;
- the spatial synchronicity bit (SSB);
- the AP_Bit.

3.3. APD FPGA Hardware Implementation

The algorithm schematized in Figure 7 has been implemented by a VHDL digital circuit, whose block diagram is shown in Figure 10. The circuit consists of a control unit (CU) and an arithmetic logic unit (ALU).

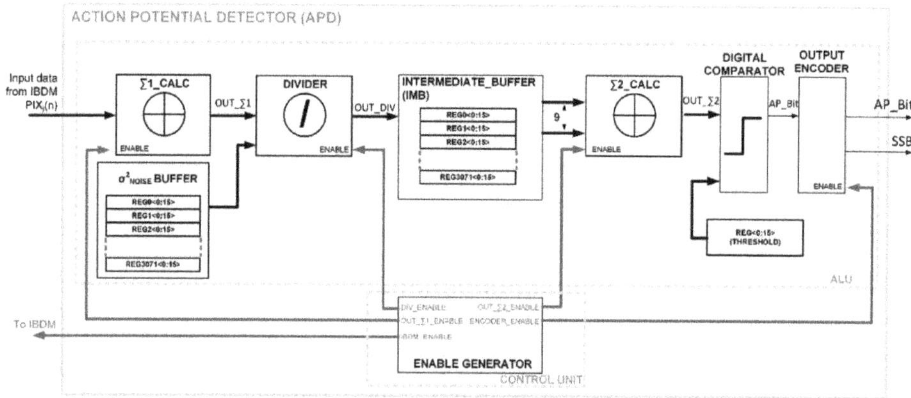

Figure 10. Action potential detector (APD) block scheme.

The ALU consists of arithmetic, logic, and memory blocks that perform the operations given in Equations (4)–(7). Input data are encoded as signed 14-bit integers, but thanks to their limited dynamic range (cantered around zero and with maximum values around 150–200), can be re-coded with nine-bit resolution (−256 to 255) to limit the hardware requirement without introducing any error. The operations are then performed with unsigned integer arithmetic since after calculating the square value in Σ1_CALC all the values become positive. After each block of the ALU, data are again re-coded to be represented with the minimum resolution compatible with their dynamic range, thus minimizing hardware requirements. The only operation that introduces computational errors while performed with integers is the division, implemented with a Radix-2 algorithm and neglecting the reminder. To minimize this error to below 1%, the numerator (*OUT_*Σ1) has been multiplied by 256 by left-shifting before performing the division. The 256-fold increase in the *OUT_DIV* value has then been compensated by increasing the *AP_Threshold* value by 256 as well. The CU regulates the behavior of the ALU using specific enable signals and spatial pointers. The enable signals sequentially activate the stages of the ALU, controlling the evolution of the algorithm. The time evolution of the CU output signals is illustrated in Figure 11. When the IBDM_ENABLE signal is high, new data is read from the input buffer in IBDM.

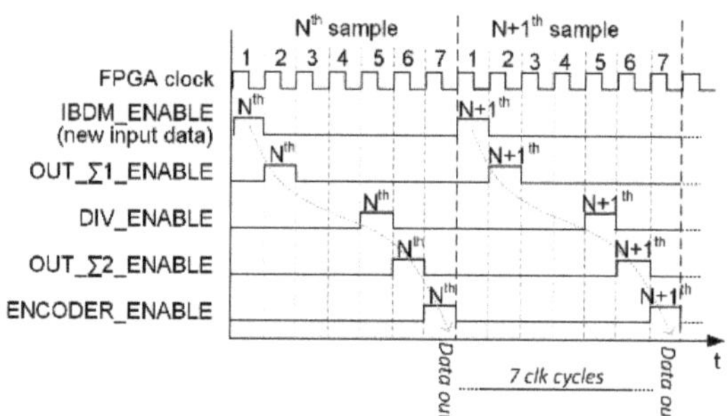

Figure 11. APD controller unit enable signals time diagram.

Afterwards the OUT_Σ1_ENABLE, the DIV_ENABLE, and the OUT_Σ2_ENABLE signals are progressively and synchronously activated. The last signal is the ENCODER_ENABLE whose main scope is to regulate the timing of the output signals (AP_Bit and SSB).

Thus, this circuit on FPGA achieves the fundamental objective to encode the entire spatial and temporal activity of the biological neuronal net in three single bits: AP_Bit (that gives the detection information), the FPGA master clock (temporal pointer), and the SSB (spatial pointer).

Each operation is performed during a single cycle of the FPGA master clock, except the division that lasts three clock cycles. Hence, to process a single data, seven clock cycles will be needed in principle. In order to reduce the required clock frequency of the hardware implementation, this design adopts a dedicated pipeline control of the ALU operation as follows. The most relevant FPGA resources utilized for this design are represented in Table 1.

Table 1. FPGA resources utilization summary.

Resource	Utilization
Number of Slice Registers	1009 (1%)
Number of Slice LUT	3278 (12%)
Number of DSP48A1s	1 (1%)
Number of RAMB16BWERs	58 (50%)

3.4. APD Control Unit Pipeline Approach

With reference to Equation (3), for a 14.1 MSample/s input data rate, the minimum frequency of the FPGA master clock for a proper control unit operation is 98.7 MHz.

It is possible to achieve the same data throughput with reduced FPGA master clock frequency by using a pipeline approach. The operations in Equations (4)–(7) are scheduled in consecutive clock cycles and the results are stored in separate registers.

Thus, these operations can be executed in parallel, saving time and reducing clock frequency. In fact, when a certain sample (N-th sample in Figures 11 and 12) is processed by a specific stage (Σ1_CALC, DIVIDER, etc.) then such stage waits for six clock cycles for the next data (($N + 1$)-th) without performing an operation and practically entering in stand-by mode.

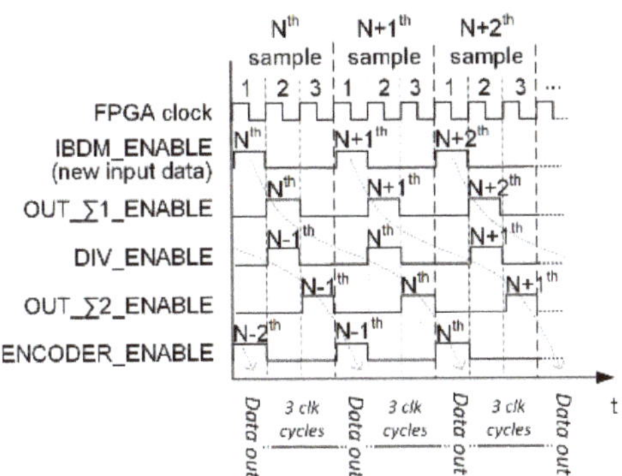

Figure 12. Pipelined controller unit enable signals time diagram.

This intrinsic inefficiency can be mitigated by providing a new data input to the ALU as soon as the first stage (Σ1_CALC) has produced an output.

In this way, the system will accept a new input data every three master clock cycles. Thus, the minimum master clock frequency to handle 14.1 MS/s input data rate will be 42.3 MHz. Figure 12 shows the CU enable time diagram used in the pipelined APD (P-APD). Each enable is now high once every three clock cycles instead of seven. Figure 13 shows the output data stream for the pipeline APD. The signal in the second line shows the input data rate, which now requires only three clock cycles to perform all the PCA operations. The final clock frequency adopted in this design is thus 42 MHz.

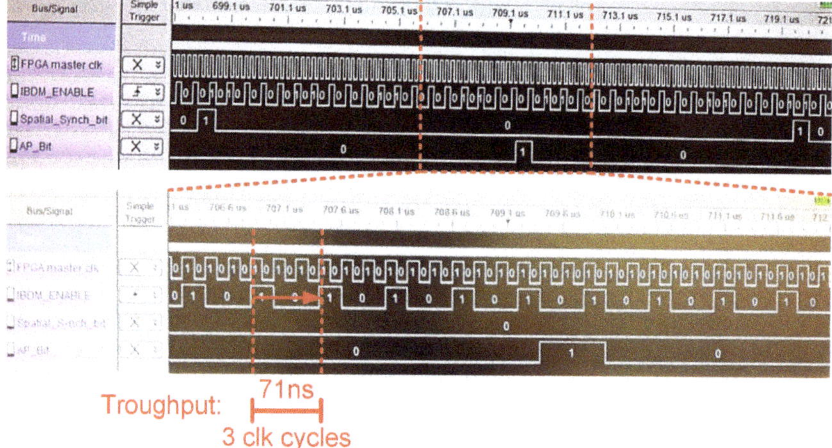

Figure 13. Pipelined action potential detector output bits.

4. Experimental Results

The NSDD for real-time detection of the neural spikes has been validated by two different setups: behavioral and biological. The behavioral setup is a dedicated test-bench schematized in Figure 14 with the main aim to check the efficacy of the digital system vs. single pixel SNR. It is based on dedicated constructed pattern with a priori known SNR. This test can verify the false positive/negative AP events and quantifying the single pixel SNR range in which the NSDD is able to operate at minimum detection faults.

On the other hand, the biological setup is used in order to directly check the NSDD behavior under the signals coming from the CMOS MEA. This setup has the ultimate scope to detect AP signals under the real-life conditions and manage the effective noise spatial map of the MEA.

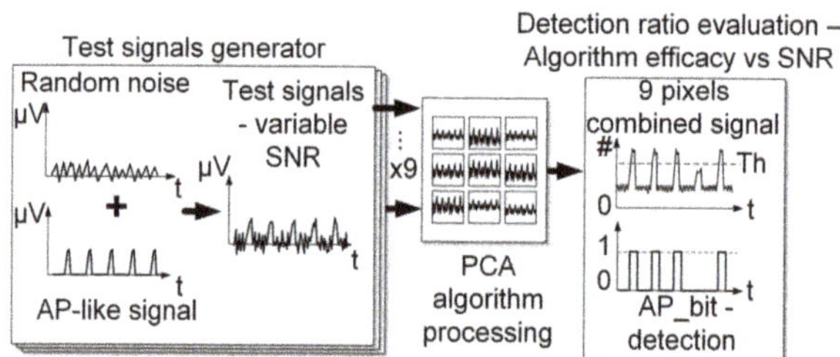

Figure 14. Behavioral setup functional scheme.

4.1. Behavioral Validation

The behavioral validation has been performed at two distinct levels: nine-pixel subset and the whole matrix.

A specific nine-signal pattern has been generated by MATLAB, where each pixel signal has a rate of 4.6 KSample/s and is composed by two main contributions: the noise and 250 AP-like signals spaced of 10 ms with 1 ms time-width.

The duration of each track is 2.5 s. Each spike in the subset has the same phase. The noise power is also the same in each pixel (120 μV_{RMS}) whereas the SNR ranges from 0 dB up to 15 dB (i.e., the last SNR corresponding to a sinusoidal signal at 5·σ power). This way, since the pattern (and thus the instant in which the AP-like signal will occur) is a priori known, it is then possible to compare the

NSDD output bit with the input pattern and to evaluate the robustness of the system (i.e., the eventual fault vs. SNR).

Figure 15 shows the percent value of the detected AP vs. the SNR. Obviously for high SNR (>7 dB) practically all the input APs will be detected by the hardware. At very low SNR the correlation algorithm features a certain detection fault (25% of detected APs at 3 dB SNR), losing several spikes.

Increasing the SNR of the single pixel up to 6 dB the NSDD has very high performance and detects 98% of APs, coherently with the error probability imposed in Equation (1).

APs with 250 $\mu V_{0\text{-peak}}$ amplitude are detected with 98% efficacy.

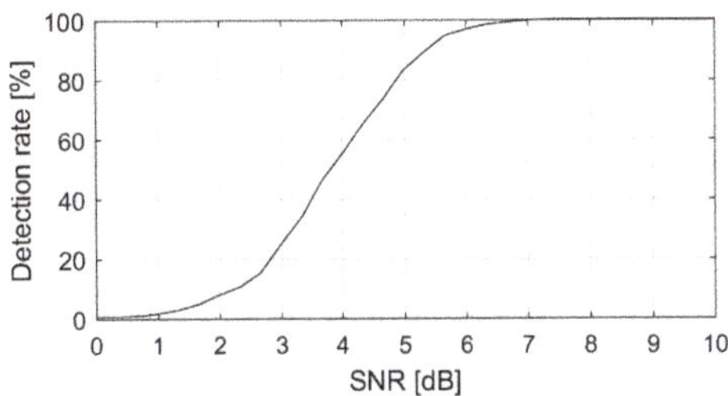

Figure 15. Percentage of detected AP versus SNR.

4.2. NSDD Biological Validation

The NSDD has been finally tested with the signals coming from the neurons culture. The hereby described system outputs in real-time (with 165 ns delay) a binary signal (AP_Bit) encoding spatial and temporal information about the detected neural spikes. This single-bit signal has two important features:

- it enables event-driven communication, and/or it can be used for instantaneous control of the electrical stimulation signal;
- it can be further processed by dedicated hardware and/or software algorithms in order to perform spatial and temporal mapping of the APs above the MEA.

More specifically, the NSDD system has been tested by monitoring the neurons culture for 15 s and using a dedicated set of MATLAB functions to extract AP information from the real-time generated AP_Bit.

This way, the data coming from FPGA-APD allows representing the neuronal cells culture electrical activity by:

- neural spike spatial map, representing the cumulative number of AP vs. pixel for a given 15 s acquisition time;
- noise power spatial map, showing the noise power spatial distribution within 2.5 s single pixel time evolution;
- action potential bursting map, detecting eventual synchronous spikes activity over the entire MEA covered area (time window is again 2.5 s).

Figure 16 shows the neural spike spatial map where colour intensity is related to the local neuron activity detected by the NSDD in 15 s time window. Notice that yellow/red regions point out 6/12 spikes over 15 s, resulting in the highest AP rate over the given time window. It has been shown that hippocampal neuronal cell cultures exhibit spontaneous electrical activity with rates coherent with the ones observed by the NSDD [10]. For sake of completeness, Figure 16 also shows the neural spike map superimposed on the culture photo.

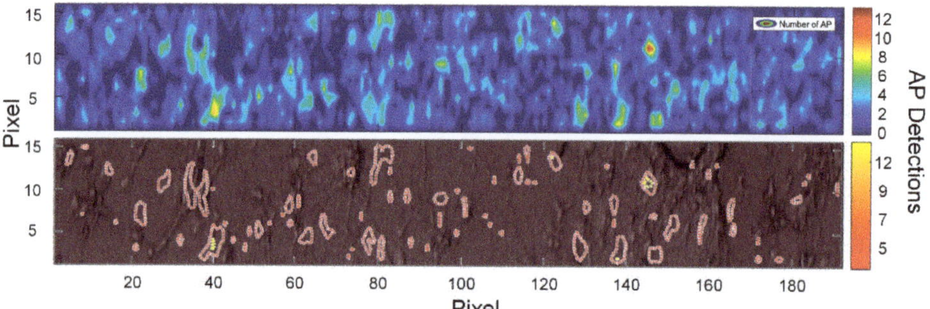

Figure 16. Neural spike (action potential) spatial map (over 15 s acquisition time).

4.3. Action Potential Bursting

The AP_Bit can be used to detect whether any synchronous activity of the neuron population is happening at any given moment. Simultaneous spiking by multiple neurons could hint at particular medical conditions such as the beginning of an epilepsy seizure. In order to produce strong synchronous behaviours in cell cultures, external chemical/electrical stimulations are usually required [16,17]. However, for this experiment no external stimulus has been provided, and therefore only minor synchronicity levels have been observed. More precisely, by counting the total APs detected over 10 ms time-windows, few points over MEA area have shown at some point some (casual/random) synchronous spiking. Figure 17 shows the total number of synchronous spikes referred to the whole MEA during a 10 ms time window. It is possible to observe that some pixels have disclosed some synchronous spiking, for example between 910 ms and 920 ms a total of four areas on the MEA produced some APs. The correspondent spatial map for this particular bursting event is also shown in the bottom part of the same Figure 17. This demonstrates the capability of the proposed system to operate in complex cultures where a more pervasive synchronous activity is imposed. In fact, the NSDD minimizes the data volume produced by the MEA by efficiently encoding only the relevant information about actual neuron activity. This greatly simplifies the real-time implementation of subsequent higher-level algorithms such as AP bursting detection or, more generally, algorithms that can detect neural patterns and abnormal neural activity (i.e., epileptic seizures).

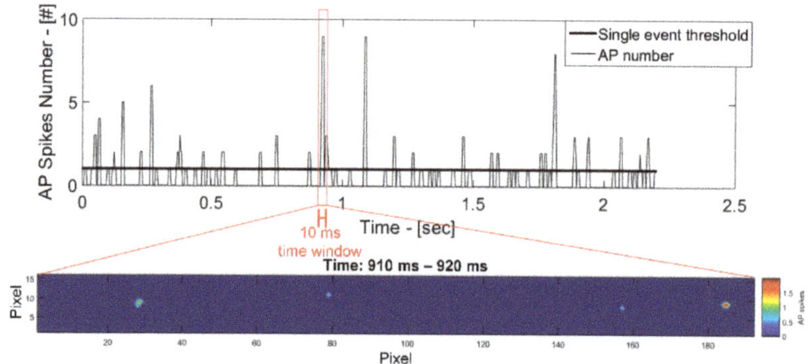

Figure 17. Neuronal spikes map of action potential bursting.

5. Conclusions

This paper presents the design of a complete FPGA-based digital circuit that monitors the electrical activity of a hippocampal neuronal cells culture over a micro-electrode array and detects action potentials from background noise. The system has been almost entirely implemented by dedicated hardware solutions that allow real-time detection of the on-going neural activity and encodes the temporal and spatial information using a dual bit stream at 14.1 MHz. The system is composed by the VHDL-based FPGA design and a proper set of MATLAB functions for data evaluation. A behavioral validation testbench has been developed and has been used to extrapolate the algorithm detection efficacy vs. SNR.

The system has been experimentally validated by producing a spatial map of the on-going electrical activity, which was shown consistent with spontaneous neural activity rates. The hereby proposed system can also provide the spatial distribution of the noise power over the biosensor, which is linked to adhesion/presence of cells over the MEA surface. Finally, the output bit-stream has been used to detect AP bursting events, highlighting synchronous neural activity over the biosensor.

Author Contributions: M.D.M. and S.V. conceived the experiments; E.A.V. designed the FPGA-based system and performed the experiments in collaboration with M.R., D.G., F.R. and G.C.; A.B. supervised the digital design; M.M. prepared the neuronal cells cultures; R.Z. provided technical support; All the authors contributed to the writing of the paper.

Funding: This research received no external funding.

Conflicts of Interest: The authors declare no conflict of interest.

References

1. Rutten, W.L. Selective electrical interfaces with the nervous system. *Ann. Rev. Biomed. Eng.* **2002**, *4*, 407–452. [CrossRef] [PubMed]

2. Hochberg, L.R.; Serruya, M.D.; Friehs, G.M.; Mukand, J.A.; Saleh, M.; Caplan, A.H.; Branner, A.; Chen, D.; Penn, R.D.; Donoghue, J.P. Neuronal ensemble control of prosthetic devices by a human with tetraplegia. *Nature* **2006**, *442*, 164–171. [CrossRef] [PubMed]

3. Vallicelli, E.A.; De Matteis, M.; Baschirotto, A.; Rescati, M.; Reato, M.; Maschietto, M.; Vassanelli, S.; Guarrera, D.; Collazuol, G.; Zeiter, R. Neural spikes digital detector/sorting on FPGA. In Proceedings of the Biomedical Circuits and Systems Conference (BioCAS), Turin, Italy, 19–21 October 2017.

4. Vallicelli, E.A.; Fary, F.; Baschirotto, A.; De Matteis, M.; Reato, M.; Maschietto, M.; Rocchi, F.; Vassanelli, S.; Guarrera, D.; Collazuol, G.; et al. Real-time digital implementation of a principal component analysis algorithm for neurons spike detection. In Proceedings of the 2018 International Conference on IC Design & Technology (ICICDT), Otranto, Italy, 4–6 June 2018.

5. Kim, R.; Joo, S.; Jung, H.; Hong, N.; Nam, Y. Recent trends in microelectrode array technology for in vitro neural interface platform. *Biomed. Eng. Lett.* **2014**, *4*, 129–141. [CrossRef]

6. Cogan, S.F. Neural stimulation and recording electrodes. *Annu. Rev. Biomed. Eng.* **2008**, *10*, 275–309. [CrossRef] [PubMed]

7. Kim, S.; Bhandari, R.; Klein, M.; Negi, S.; Rieth, L.; Tathireddy, P.; Toepper, M.; Oppermann, H.; Solzbacher, F. Integrated wireless neural interface based on the Utah electrode array. *Biomed. Microdevices* **2009**, *11*, 453–466. [CrossRef] [PubMed]

8. Eversmann, B.; Jenkner, M.; Hofmann, F.; Paulus, C.; Brederlow, R.; Holzapfl, B.; Fromherz, P.; Merz, M.; Brenner, M.; Schreiter, M.; et al. A 128x128 CMOS biosensor array for extracellular recording of neural activity. *IEEE JSSC* **2003**, *38*, 2306–2317.

9. Meister, M.; Pine, J.; Baylor, D.A. Multi-neuronal signals from the retina: Acquisition and analysis. *J. Neurosci. Methods* **1994**, *51*, 95–106. [CrossRef]

10. Maschietto, M.; Girardi, S.; Dal Maschio, M.; Scorzeto, M.; Vassanelli, S. Sodium channel β2 subunit promotes filopodia-like processes and expansion of the dendritic tree in developing rat hippocampal neurons. *Front. Cell. Neurosci.* **2013**, *7*, 2. [CrossRef] [PubMed]

11. Lambacher, A.; Vitzthum, V.; Zeitler, R.; Eickenscheidt, M.; Eversmann, B.; Thewes, R.; Fromherz, P. Identifying firing mammalian neurons in networks with high-resolution multi-transistor array (MTA). *Appl. Phys. A Mater. Sci. Process.* **2011**, *102*, 1–11. [CrossRef]

12. Nicolaou, N.; Constandinou, T.G. A Nonlinear Causality Estimator Based on Non-Parametric Multiplicative Regression. *Front. Neuroinform.* **2016**, *10*, 19. [CrossRef] [PubMed]

13. Available online: https://www.xilinx.com/products/design-tools/coregen.html (accessed on 2 December 2018).

14. Available online: http://www.venneos.com/product/can-q-analyzer.html (accessed on 2 December 2018).

15. Available online: https://www.agilent.com/home (accessed on 2 December 2018).

16. Salam, M.T.; Velazquez, J.L.P.; Genov, R. Seizure suppression efficacy of closed-loop versus open-loop deep brain stimulation in a rodent model of epilepsy. *IEEE Trans. Neural Syst. Rehabil. Eng.* **2016**, *24*, 710–719. [CrossRef] [PubMed]

17. Biffi, E.; Regalia, G.; Menegon, A.; Ferrigno, G.; Pedrocchi, A. The influence of neuronal density and maturation on network activity of hippocampal cell cultures: A methodological study. *PLoS ONE* **2013**, *8*, e83899. [CrossRef] [PubMed]

 *electronics*

Review

CMOS Interfaces for Internet-of-Wearables Electrochemical Sensors: Trends and Challenges

Michele Dei [1,*], Joan Aymerich [1], Massimo Piotto [2], Paolo Bruschi [2], Francisco Javier del Campo [1] and Francesc Serra-Graells [1,3]

[1]   Institut de Microelectrònica de Barcelona IMB-CNM (CSIC), 08193 Barcelona, Spain;
      joan.aymerich@imb-cnm.csic.es (J.A.); javier.delcampo@csic.es (F.J.d.C.);
      paco.serra@imb-cnm.csic.es (F.S.-G.)
[2]   Dipartamento di Ingegneria dell'Informazione of University of Pisa, 56126 Pisa, Italy;
      massimo.piotto@unipi.it (M.P.); paolo.bruschi@unipi.it (P.B.)
[3]   Department of Microelectronics and Electronic Systems, Universitat Autònoma de Barcelona,
      08193 Barcelona, Spain
*   Correspondence: michele.dei@imb-cnm.csic.es; Tel.: +34-935-94-77-00 (ext. 2492)

Received: 31 December 2018; Accepted: 28 January 2019; Published: 31 January 2019

Abstract: Smart wearables, among immediate future IoT devices, are creating a huge and fast growing market that will encompass all of the next decade by merging the user with the Cloud in a easy and natural way. Biological fluids, such as sweat, tears, saliva and urine offer the possibility to access molecular-level dynamics of the body in a non-invasive way and in real time, disclosing a wide range of applications: from sports tracking to military enhancement, from healthcare to safety at work, from body hacking to augmented social interactions. The term *Internet of Wearables* (IoW) is coined here to describe IoT devices composed by flexible smart transducers conformed around the human body and able to communicate wirelessly. In addition the biochemical transducer, an IoW-ready sensor must include a paired electronic interface, which should implement specific stimulation/acquisition cycles while being extremely compact and drain power in the microwatts range. Development of an effective readout interface is a key element for the success of an IoW device and application. This review focuses on the latest efforts in the field of Complementary Metal–Oxide–Semiconductor (CMOS) interfaces for electrochemical sensors, and analyses them under the light of the challenges of the IoW: cost, portability, integrability and connectivity.

Keywords: integrated circuits design; smart wearables; Internet of Wearables; Internet of Things; CMOS Electrochemical Sensing; Flexible Technologies; Hybrid Integration Technologies

---

1. Introduction

At the onset of the era of connected intelligence, many real-life issues have been addressed through massive data collection from a community of users. Internet-of-Things (IoT) devices are already shaping our lives through augmented information, entertainment and social interactions. Recently, the IoT market has been experiencing the exponential diffusion of smart unobtrusive wearable devices aimed at determining important vital parameters of human beings. In this paper, we will discuss the term Internet of Wearables, IoW, for these extremely innovative devices distinguished from the previous generation of portable biomedical devices for their miniaturization, portability and internet-ready communication capabilities. In this scenario, miniaturized electrochemical sensing are likely to play a key role in many contexts including: military, security, food quality and safety, healthcare, wellness and environmental monitoring. The relevance of electrochemical sensing through dedicated CMOS interfaces is demonstrated by a number of recent reviews [1–3]. The ambition of this work is to provide a complete survey giving all the relevant information about the electrochemical sensing and the most advanced CMOS instrumentation circuits and techniques with an emphasis on Internet-of-Wearables (IoW) application. This article is organized as follows: Section 2 gives an overview of the latest trends in smart wearables, Section 3 summarizes the fundamentals of electrochemical sensing most commonly used in the field of wearables, Section 4 offers an extensive and detailed account of CMOS interfaces for electrochemical sensors. This work concludes by discussing the main challenges facing CMOS design for IoW.

2. Smart Wearables and the IoT: Basic Requirements and Main Fields of Application

Wearable technologies are one of the fastest growing industries of our time, expected to generate revenues over \$51 bn by 2022 at a gaping compound annual growth rate of 15% [4]. Smart wearables are electronic devices whose mission is to interface humans with the digital world [5]. The goal is to improve the quality of life or performance of the users by sensing the person wearing it and its environment, and providing these data to a processing unit able to inform or assist the user in different ways depending on the specific application.

Pioneering devices, such as activity monitors, successfully integrate multiple physical sensors. However, without complementary biochemical information, it is nearly impossible to form a complete picture of an individual's health status in real time. However, the integration of chemical sensors is more challenging compared to physical ones due to the differences between them in terms of fabrication technologies, calibration requirements, and operating lifetime. As will be discussed below, physical sensors are in many cases compatible with CMOS processes and many conventional fabrication techniques, and have lifetimes on the order of months or years. Chemical sensors, on the other hand, are incompatible with many microelectronic processes, particularly CMOS, and require frequent calibration or replacement. Consequently, huge opportunities are emerging in the area of chemical sensing, but they come hand in hand with no less significant challenges [6].

Nowadays, the development of smart wearable chemical sensors is driven by applications in four key areas: (i) health; (ii) sports; (iii) work safety; and (iv) defense and law enforcement. There are a number of analytes of common interest across these application areas, such as glucose and electrolytes, and then there are other analytes of particular relevance to a single application, as in the case of certain hormones, drugs of abuse and toxins. Regarding transduction mechanisms, most sensors rely on either optical or electrochemical sensors. This review addresses recent development in CMOS interfaces for the latter type.

2.1. Wearables and Non-Invasive Monitoring

Figure 1 depicts suitable body areas for placement of wearable chemical sensors, as well as typical form factors and application areas. On the other hand, Table 1 provides physiological concentration ranges of analytes of interest in wearable chemical sensors in body fluids suitable for the development of non-invasive detection, including wearable technologies [7]. Ideally, wearable chemical sensors should be non-invasive and, among candidate body fluids, such as sweat [6], saliva [8,9], urine and tears [10], sweat seems the most suitable medium to address biochemical parameters unobtrusively. This is reflected by the relatively higher number of works reporting sweat sensors [11–15] in the literature, compared to saliva or tears. Urine is a special case because, although it enables non-invasive determination of biochemical parameters, it does not lend itself for non-invasive monitoring applications. Depending on which of these fluids is addressed, the most common sensor forms are skin-patches, followed by contact lenses. Salivary analysis, on the other hand, is typically performed from swabs.

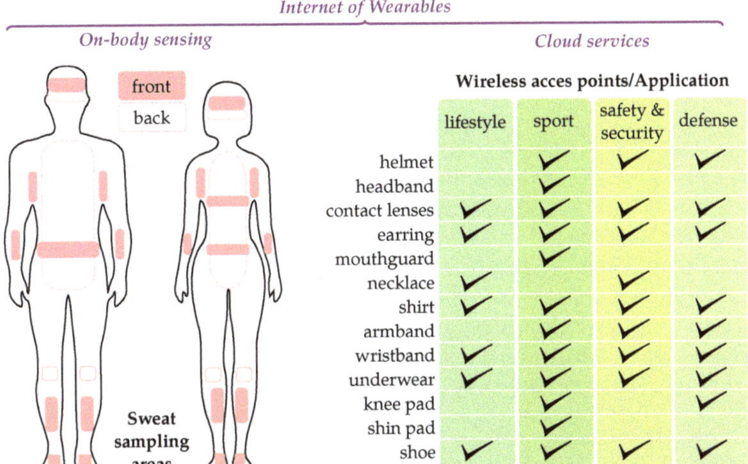

Figure 1. Internet of Wearables concept. Sweat sampling areas in the human body for in situ measurement of physiological analytes and possible wireless access points for data links.

Table 1. Normal ranges for analytes of interest in body fluids.

Body Fluids	Blood	Sweat	Saliva	Tear	Urine [a]
References	[16]	[17–20]	[8,9]	[16,21–23]	[24–27]
Glucose [mM]	3.3–6.7	0.06–0.11	0.22–0.72	0.2–1	2.78–5.5
pH	7.36–7.44	4–5.5–7	6.2–7.6	6.5–7.6 7.14–7.82	4–8
$[Na^+]$ [mM]	136–145	10–40 [b]	20–80	80–161	40–220
$[K^+]$ [mM]	3.5–5	4–5 [c]	20	not found	25–125
$[Cl^-]$ [mM]	98–107	<40	30–100	106–130	110–250
Lactate [mM]	0.3–1.3 [d] 0.36–0.75 [e]	20–60 5–110	0–0.4 [f]	unknown	0.5–2.2 [g]
IL-6 [pG/mL]	0–4.3	7–16	2.5	100–200	20–30

[a] For urine, quantities are expressed as [mmol/day]; [b] 70 mM suggests cyctic fibrosis; [c] $[K^+] > 10$ mM indicates a likely system error; [d] venous blood; [e] arterial blood. During exercise/extreme exercise up to 12–25 mM; [f] values during/after physical exercise; [g] Measured in the interval between the first morning and the previous urine sample; first morning urine sample was obtained between 6:00 a.m. and 10:00 a.m.

2.2. Wearable Sensors in Health

The pursuit of chemical sensors for non-invasive monitoring has been a truly active field of research over the past few decades [28]. Wearable sensors can be of great help in the early diagnosis and prevention of various diseases, particularly those associated with lifestyle such as diabetes mellitus and cardiovascular diseases. Its growing prevalence makes diabetes mellitus [29] perhaps the single most important application of wearable chemical sensors today. Blood glucose is measured using amperometric or coulometric biosensors and is able to quantify the oxidation of glucose using oxidase [30] or dehydrogenase enzymes [31]. While enzyme-based systems dominate, long-term stability issues associated with enzymes have motivated very active research in non-enzymatic glucose sensors [32]. Additionally, non-enzymatic glucose sensors might be easier to produce than their enzymatic counterparts, which will eventually represent an important breakthrough.

The evolution of glucose biosensors has been addressed in countless articles and excellent reviews [23,31], including wearable and non-invasive ones [33], but a summary will be given for convenience. Glucose monitors have evolved greatly [23,34] since Clark and Lyons published the first glucose biosensor back in the 1960s [35]. Blood glucose analyzers have evolved from laboratory benchtop instruments, first commercialized by Yellow Springs Instruments (YSI Inc., Yellow Springs, OH, USA) in 1975, to home appliance in the 1980's, to portable, hand-held self-monitors in the 1990s. These instruments analyze blood using enzyme biosensors, and although they have been able to work on increasingly smaller blood samples (most current systems require less than 1 µL of blood), they still require uncomfortable and painful blood extraction using a lancet. An important technology leap was made by Cygnus' (Cygnus Inc., Redwood City, CA, USA) Glucowatch in the late 1990s [36]. Glucowatch relied on two glucose biosensors to monitor glucose in sweat, following stimulation by reverse iontophoresis through pilocarpine pads. Unfortunately, issues related to calibration, quantification of sweat glucose, and skin irritation meant the end of the Glucowatch, but it was nevertheless a very significant milestone in terms of technology and product development. Since then, the industry has gone back to an intermediate stage, between invasive daily finger-pricking and non-invasive sweat measurements, to monitor blood glucose through minimally invasive monitoring in interstitial fluid. Abbott's (Abbott Laboratories, Chicago, IL, USA) Freestyle Libre [37] monitors blood glucose through interstitial fluid over a period of two weeks [38], which represents a major improvement for the lifestyle of diabetic patients.

2.3. Wearable Sensors in Sport

Chemical sensors in sport wearables are intended to ensure top-performance in athletes [39]. This may be achieved through injury prevention and early detection [40], but also through the monitoring of effort and hydration in real time through sweat, which is the obvious sample medium to monitor, given its abundance during physical exertion [41]. In spite of this, sweat measurements also require choosing an adequate body area for sampling [17] and smart sweat management strategies [42], the most common of which are based on microfluidics [43–45] or on absorbents and lateral flow membranes [15,20,46].

The prevention of injuries, particularly in elite and professional sports, is an ambitious and elusive goal. One way to achieve it is through the early detection of biomarkers of muscular damage.

Two examples of such biomarkers are creatine kinase and cytokines such as interleukin-6 (IL-6) [47]. While creatine kinase does not show in sweat, IL-6 has been reported to appear in sweat at similar concentration levels as blood [48]. Its determination requires the use of affinity-based sensors, which, in contrast to enzyme-based biosensors, are generally unsuitable for monitoring purposes. Aptasensors, a type of affinity-based sensor that relies on synthetic receptors [49], are more reversible than immunosensors and may be adapted to semi-continuous measurements. An example of this has been recently reported of label-less detection of IL-6 in artificial sweat using electrochemical impedance spectroscopy [50]. Perhaps the most accessible analytes in sweat are electrolytes, which provide direct information on hydration levels, and are measured potentiometrically using ion selective electrodes [51]. Dehydration is an important condition to determine and manage, as total body water losses greater than 2% are likely to cause performance losses in endurance sports [52]. Other analytes, such as glucose and lactate, may also be determined in sweat, using amperometric biosensors instead. However, although the correlation between sweat and blood glucose is accepted, the case of lactate, an indicator of muscular effort, is not so clear-cut. Blood lactate is typically measured in capillary blood, typically from the ear lobe [53]. Although the correlation between blood lactate and sweat lactate levels has been reported [19], and it is accepted that sweat lactate may be used as an indirect indicator of physical activity, sweat lactate is a more likely indicator of sweat gland activity itself [54].

2.4. Safety at Work

This is another area where non-invasive monitoring of biochemical parameters can help prevent accidents and injuries in physical work environments. Examples of wearables featuring physical sensors can be found in the literature [55].

In terms of chemical and biochemical information, the same parameters and sensors applied in sport could be applied to the workplace to monitor the fitness level of a pool of workers. In addition to level of hydration and effort, other risk factors such as stress and drugs of abuse may be detected non-invasively too [56]. Small sized hormones, such as serotonine, dopamine and cortisol, may be used to assess an individual's emotional state non-invasively, as they appear in sweat, saliva, tears and urine [27]. The difficulties in the detection lie in the low concentration levels of these molecules in sweat (in the nM-μM range), and the fact that detection in most cases requires immunosensors, which, as mentioned earlier, are unsuitable for continuous measurements required in a monitoring application. Barring ethanol, which is easily detected in breath and sweat electrochemically, the intake of most drugs of abuse is very hard to monitor non-invasively and impractical to embed in a wearable device because their detection involves immunoassays.

2.5. Defense and Law Enforcement

This is another area of great interest, which can also benefit from the same technologies described above for sports. However, these applications normally have much more stringent requirements in terms of reliability and robustness. This makes them particularly challenging to develop, but, correspondingly, highly valuable and attractive in economic terms. Recent examples in this area involve wearable sensors for the detection of nerve agents [57–59].

2.6. Design and Fabrication of Wearable Devices

Application scenarios define the functional requirements of wearable devices. This section aims to highlight a few important ideas that need to be considered in relation to the design of wearable sensors. The form factor of a wearable device defines aspects such as its physical parts, their shape and size. Most wearable devices known today, such as activity monitors and smartwatches, are "one-part" devices because they rely on physical sensors alone. This is possible because those physical sensors will last as long as the device itself—whose lifetime is usually limited by its power source—and because they are produced using Microelectromechanical systems (MEMS) technologies that are fully or highly compatible with the CMOS processes employed in the fabrication of microprocessing units. Therefore, it is relatively easy to integrate them inside or along other silicon components, relying on System-on-Chip, SoC, technologies vs. System-in-package, SiP, and heterogeneous integration processes [60,61].

Chemical sensors, on the other hand, are much shorter lived than any other component in these smart microsystems. The (bio)sensor recognition element determines the longevity. Thus, ignoring calibration, ion selective electrodes may operate for weeks, enzyme biosensors can last hours to days, and affinity-based biosensors, such as immunosensors, may just be good for a single use or a few hours. Although different approaches have been taken to overcome their operational stability problems, the most common one consists in separating the sensor from the electronics, as in the case of present-day glucometers, where the biosensing part is a disposable strip that is inserted inside a reader. The same model was applied to the Glucowatch, and only Abbott's Freestyle Libre departs from it by integrating the biosensor into a disposable microsystem that interfaces wirelessly with the reader. In this last case, there is a real advantage if the benefits for the user outweigh the (economic and environmental) costs of disposing of a button battery, a printed circuit board, PCB, containing several discrete components, and a plastic casing. This approach of integrating discrete components in a disposable package has also been adopted in several works present in the literature. However, even if the components can be produced inexpensively, the cost of heterogeneously integrating them in flexible or elastic substrates, is still too high [62–64].

We can see only two solutions to this problem until the cost of integration becomes affordable. One is to reduce the overall number of silicon components to a single chip, and with as few contact pads as possible. The other solution involves devices that do not require silicon or any kind of discrete components at all [65]. This second approach may have been completely unfeasible only a few years ago, but progress in printed electronics may soon enough make it possible [66–68].

In both cases, interfaces between the wearable and the user and between the wearable and the rest of the world needs to be carefully studied. The first one defines important aspects of usability and comfort, while the second set of interfaces determines how a disposable sensor interacts with its control instrumentation. The physical connection required between electrochemical sensors and their control instrumentation, namely a potentiostat, represents a significant hurdle in a wearable system. In order to avoid wiring and connectors, which would increment the complexity and the cost of heterogenous solutions, recent works focused on integration of the CMOS die along sensor and data-linked wirelessly via Near-Field Communication (NFC) or other short-range radio-frequency (RF) signals [14,69,70]. Works reporting wireless links are still a minority, but they are growing rapidly [71]. RF can and should be used not only to extract information from the device, but also to power it. This can alleviate the need for on-board power sources and their associated control components, which take up precious space and can make the final device not only bulkier, but also much more costly.

2.7. Data-Access Points: Smartphone, Smartwatch and WPAN Radio

The next generation of smart wearables will adhere to the IoT paradigm embodying user-centred Cloud-connectedness devices. The user will seek control and interaction with data generated from its body and/or its surroundings without the burden of heavy and bulky devices. Connection to the Cloud is also fundamental to enrich the wearable experience through ambient intelligence. In order to emphasize this latter concept, we coin here the term Internet of Wearables. The IoW concept ecompasses a number of aspects which go beyond the scope of this review. Nevertheless, we find it useful to employ this term since it entails specific characteristics for the IoW interface circuits as will be clear in the following discussion.

Smartphones currently play a crucial role in the definition of the smart wearables, since they are the most popular and widespread IoT platforms. This has been the subject of a number of dedicated reviews due to their amenability as portable instrument for biosensing and self-diagnostics [72–75]. The importance of smartphones resides in the combination of the following factors: (i) they already carry powerful data processing hardware; (ii) they enable direct and intuitive user interfacing; (iii) they allow for Cloud connectivity and (iv) they provide a number of data-access ports including camera, USB, NFC and audio-jack that can be exploited to interface with the smart wearables.

Smartwatches are also gaining popularity as powerful e-gadgets and have also been a subject of interest for researchers as a tool to gather biometric data [76,77]. Their body conformability and low weight represent an important step ahead towards the IoW.

IoW devices may communicate each other directly through medium/short-range wireless connectivity or through another device, e.g., a smartphone. In both cases, radio capabilities must be embedded. In order to enable interoperability between the IoW devices and their surroundings, compliance with wireless standards is a compulsory need. In this sense, a number of standards from the IEEE 802.15 working group and from the International Organization for Standardization are of interest for IoW applications.

Table 2 lists the trends in the state-of-the-art research on smart wearables equipped wireless off-the-shelf components. It emerges that wireless personal area networks, WPANs, such as Zigbee and Bluetooth are currently employed for medium/short range communications. The Bluetooth Low Energy variant, is particularly attractive for its smart power management that allows for a long battery life. An IoW-ready CMOS interface must be capable of communicating with the transceiver chip through the provided digital I/O interfaces, usually comprising Serial Peripheral Interface (SPI) and/or I2C protocols.

For totally passive, i.e., battery-less, operation, the NFC technology is employed. In this case, the IoW interface must include all the RF front-end circuits (rectifier, load modulator, limiter, low-dropout regulator, envelope detector, demodulator and clock recovery circuit) and the digital control logic needed to implement the specific NFC ISO/IEC protocol used for the application. The great advantage of getting rid of the burden of the battery comes at the cost of limited wearability, since the smartphone or other NFC-capable device needs to be kept aligned with the sensing tag during the measurement cycles, which, in the case for electro-chemical sensing, may extend in the range of minutes.

New possibilities may emerge in the immediate future from research in bio-fuel cell sensing, where both analytical information and operational power can be extracted from the electrochemical interface. Recently, authors in [78] demonstrated that the power harvested from an electronic skin bio-fuel cell is high enough to operate a conventional Bluetooth Low Energy (BLE) radio chip. Section 4.5 reports the last pioneering custom integrated circuits interfaces in this field.

Table 2. Wireless personal area networks-radio options based on off-the-shelf components for Internet of Wearables.

Radio	Off-The-Shelf Component(s): Radio + MCU	Reference(s)
Zigbee (IEEE 802.15.4)	Xbee-PRO RF + PIC24HJ12GP201	[79]
	Xbee + Arduino Lilypad	[80]
	nRF2401 + ATMega128L	[81]
	CC2530	[82]
Bluetooth Low Energy (IEEE 802.15.1)	CC2541	[83–85]
	HC-06 + ATMega328p	[43,64,86]
Bluetooth 2.0	CC2560 + MSP430BT5190	[87]
NFC/RFID (ISO 15693)	Commercial tag, custom modified	[88]
	SL13A	[89]
	MLX90129	[62,90]

3. Electrochemical Sensing Overview

Electroanalytical methods provide specific information on chemical and biochemical systems by looking at the interplay between chemical species and electrical quantities at the interface between the sensor and the test solution.

When an electrode is immersed in a solution, a charge separation occurs at the interface, and an electrical double layer is established. The charge developed at the electrode surface is balanced by the solution across a small volume consisting of both adsorbed ions and ions in solution. Overall, this volume extends over a distance of a few nanometers. This is where (most of) the electrode potential is developed and electron transfer between the electrode and species in solution takes place [91,92].

The electrochemical nature of the electrode-solution interface can be used to develop (electro)analytical methods reliant on the relation between electric current, potential or impedance and the concentration of a target analyte [93]. Thus, electroanalytical methods are classified according to the measured electrical quantity as: (i) potentiometric; (ii) amperometric; and (iii) impedance-based.

3.1. Potentiometric Sensing

The impossibility to measure the potential drop across a single electrode–solution interface determines that two electrodes are required to carry out potentiometric measurements.

As depicted in Figure 2, these two electrodes are: an indicator electrode, i.e., an ion-selective electrode (ISE), and a reference electrode, RE, whose potential is stable and independent of the sample composition. In order to ensure the stability of the reference electrode potential, an important feature of potentiometric measurements is that they are carried out on an open circuit. Thus, any potential change measured across this two-electrode cell can be exclusively attributed to the indicator electrode–solution interface, and related to analyte concentration changes through an expression of the kind:

$$V_{cell} = \text{constant} + \frac{RT}{nF} \ln[\text{analyte}]. \tag{1}$$

In Equation (1), R is the universal gas constant, T is the temperature, F is the Faraday's constant and n is the valency of the ion. In the following discussion, we will use the known equivalence expression containing the Boltzmann constant K and the fundamental electron charge q:

$$\frac{RT}{F} = \frac{KT}{q} = U_T, \tag{2}$$

where U_T is the thermal voltage, according to a definition inherited from the semiconductor physics terminology.

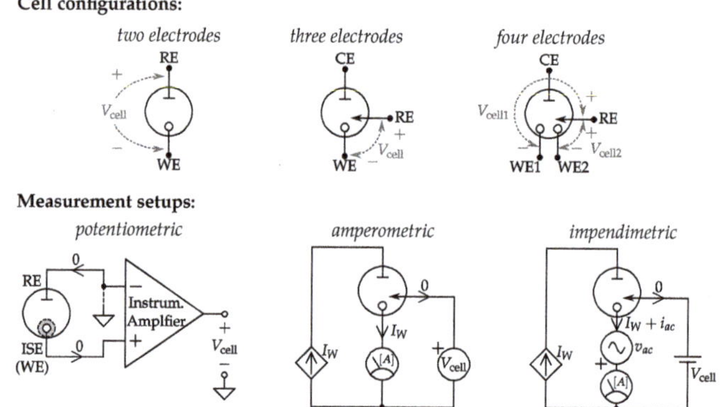

Figure 2. Electrochemical transducer cell typical configurations and measurement setups.

In an ion-selective electrode [94], the indicator electrode presents a membrane containing an ionophore that is selective to the target ion. However, in practice, other ions may also permeate through the membrane and affect the sensor response.

The typical response at an ion selective electrode is depicted in Figure 3. At low analyte concentrations, the sensor response is determined by the concentration of the interfering ions. Above a certain concentration, i.e., the limit of detection, the sensor response is linearly dependent on the logarithm of the analyte concentration. In the case of singly charged species, a slope of $U_T / \log_{10} e \approx 59\,\text{mV}$ per decade (of concentration at room temperature) is referred to as Nernstian, and it can be used to make a quick estimate of the resolution and sensitivity required in the instrumentation for a particular application.

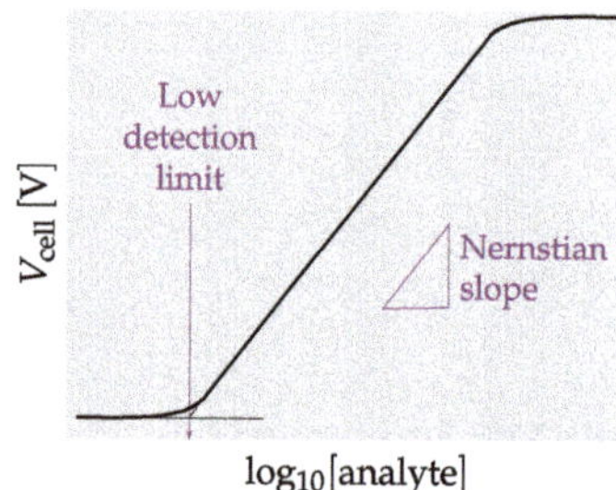

Figure 3. Schematic Ion-Selective electrode (ISE) static characteristic.

3.2. Amperometric Sensing

Amperometric techniques are based on the measurement of the current response at a given electrode potential. In contrast to potentiometric sensors, which only require two electrodes, amperometric techniques require three electrodes, namely working (sensor), reference, and auxiliary electrodes, indicated in Figure 2 as WE, RE and CE, respectively. The role of the RE in amperometry is to control the potential at the WE. To enable this control, no current can flow through the RE, and hence the need for a third, auxiliary electrode whose function is to supply the current required by the WE, I_W. Since the CE is typically larger than the WE, the process occurring at CE does not limit I_W.

Electrode processes are generally represented by an equation such as:

$$O + ne \underset{k_b}{\overset{k_f}{\rightleftharpoons}} R, \tag{3}$$

where O and R are the respective oxidized and the reduced form of a species, e is the transferred electron and k_f and k_b are the reduction and oxidation rates, respectively. Equation (3) describes a general redox process, which is also referred to faradaic process, since it is governed by Faraday's law stating that the amount of chemical reaction (mass) caused by the flow of current is proportional to the total charge passed through the electrode/solution interface. Rate constants k_f, k_b are potential dependent:

$$k_f(V_{\text{cell}}) = k_0 \exp\left(-\alpha\frac{V_{\text{cell}} - U_0}{U_T}\right); \qquad k_b(V_{\text{cell}}) = k_0 \exp\left((1 - \alpha)\frac{V_{\text{cell}} - U_0}{U_T}\right). \tag{4}$$

When V_{cell} equals the standard potential U_0, both k_f, k_b are equal to k_0 which is known as the standard heterogeneous rate constant. The standard potential, U_0, defines the activation threshold of the redox process, while k_0 measures the kinetic facility of the redox couple. The charge transfer coefficient α indicates the symmetry of the electron-transfer process. It is commonly assumed to be around 0.5, although in most cases it is found to be between 0.3 and 0.7. The k_f, k_b rate constants are controlled by the voltage drop across the electrical double layer, which in turn depends on the potential applied between the working electrode, WE, and the RE.

Notation in Equation (4) differs slightly from classical notations used in Electrochemistry books with the purpose to highlight similarities with the bipolar transistor and diode models which are more familiar in the field of electronic engineering.

The faradaic process is then governed by the current-overpotential equation:

$$I_{W,\text{faradaic}} = i_0\left(\frac{c_{O,S}}{c_O^*}\exp\left(-\alpha\frac{V_{\text{cell}} - U_0}{U_T}\right) - \frac{c_{R,S}}{c_R^*}\exp\left((1 - \alpha)\frac{V_{\text{cell}} - U_0}{U_T}\right)\right), \tag{5}$$

$$i_0 = FAk_0(c_O^*)^{(1-\alpha)}(c_R^*)^{\alpha},$$

where $c_{O,S}$ and $c_{R,S}$ are the O and R concentrations at the WE, respectively, while c_O^* and c_R^* indicate the same concentrations in the bulk (far from WE). In dynamic conditions, $c_{O,S}$ and $c_{R,S}$ differ from their respective values in bulk and obey Fick's second law of diffusion. While analytical treatment is out of scope in the context of this review, details can be found in general Electrochemistry books [91,92]. However, Equation (5) already carries all the valuable information for sensing purposes: i_0 depends on the concentration of a target analyte, which may be whether O or R. Dynamic experiments, which involve temporal transients, also holds electro-analytical information through the c_S/c^* ratios, which will be apparent in the following discussion.

Oxidation currents mean that electrons pass from the solution to the anode and, conversely, reduction currents involve the passage of electrons from the cathode to the solution. The CE completes the current path. At the same time, the current is sustained by ions in the electrolyte: in this region, the voltage drop between the working and the reference electrode, also known as ohmic, or iR-drop, should be minimized to ensure control over the working potential.

The current passing through the working electrode corresponds to the oxidation or reduction of the target analyte at the measurement potential. This measurement potential is chosen so that the current is not limited by the electrode kinetics, but mass-transport controlled. Analyte mass transport to the WE is influenced by diffusion, convection and migration. Amperometric experiments are usually designed for diffusion to prevail over the other transport mechanisms. In this way, temperature/density gradients and electrostatic forces do not influence I_W. Mass-transport control requires the application of a certain overpotential, which is the difference between the applied potential and the formal potential of the analyte redox process, $V_{cell} - U_0'$. The formal potential U_0' differs from the previously defined standard potential U_0 in that U_0' takes into account non-standard conditions, i.e., $T \neq 298\,\text{K}$, $pH \neq 0$, $[\text{analyte}] \neq 1\,\text{M}$ and how the medium or the presence of other electrolytes affects the activity of the analyte. While the applied potential should ensure mass transport control, it also ought to be low enough to prevent the oxidation or reduction of other electroactive species present in solution, and which may result in determination errors.

Controlled V_{cell} potential electroanalytical methods include potential step and linear sweep-based techniques, such as cyclic voltammetry [91–93]. A key aspect of these techniques is that the initial potential must be such that no overall faradaic current is observed. In other words, the initial potential should be close to the open circuit potential. This facilitates that the current measured during the experiment is due to the main process of interest alone. Figure 4 represents the potential functions and current response corresponding to a potential step as well as a cyclic voltammogram. In a simple potential step experiment, the transient current at the WE is monitored while V_{cell} is stepped to the measurement potential, as sketched in Figure 4a. During the first part of the experiment, the current corresponds to the charging or discharging of the electrical double layer. Depending on the electrode size, material and structure, this charging current may last for up to several milliseconds, during which the underlying faradaic current is much weaker and hard to detect. Once the double layer is discharged, the remaining current corresponds to faradaic processes, and can be measured more safely. Advanced potential step techniques, such as square wave and differential pulse voltammetry, are designed to eliminate the charging current contribution and enhance the faradaic component, but they require relatively fast instrumentation and therefore their application in wearables is not very attractive.

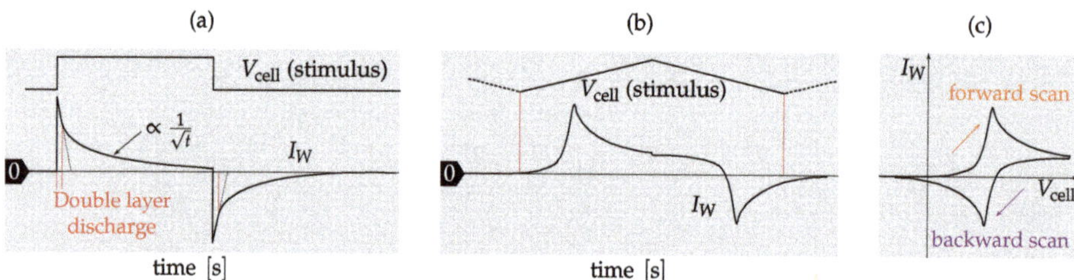

Figure 4. Electrochemical experiments: (**a**) chronoamperometry (versus potential step); (**b**) chronoamperometry (versus potential scan); (**c**) voltammogram.

Another well-established electrochemical method is cyclic voltammetry. In this case, a potential ramp is applied to the WE instead of a step, and the current response is recorded, as indicated in Figure 4b. At some point in the potential scale, well past U_0', the scan direction is reversed. This results in the recording of a forward and a backward current scan (Figure 4c). Analysis of these current responses provides rich and extremely useful information. Peak currents and positions may be used to establish kinetic and thermodynamic information about the system at hand, and the scan rate provides access to additional information in the time domain.

Although most amperometric systems are based on three-electrode cells, occasionally we may also find four-electrode detection systems based on two working electrodes that can be independently biased. This is also schematized in Figure 2. These systems may be used either for the detection of two independent analytes or in signal amplification strategies known as generator–collector systems [95–98]. In these systems, which require the analyte to be electrochemically reversible, an electrode is biased above and the other below the formal potential of the analyte. Thus, the oxidized and reduced species shuttle between electrodes, enhancing the measured current. Additionally, because many common analytes are irreversibly oxidized or reduced, using a redox cycling approach minimizes the contribution of these interfering species which cannot be regenerated [99].

3.3. Electrochemical Impedance Spectroscopy

Potentiometry and amperometry are direct current, DC, techniques. Electrochemical impedance spectroscopy, EIS, on the other hand, is based on the application of alternate signal, typically potential, across the working electrode [100]. The corresponding current response allows the calculation of the system impedance, which is a complex quantity with a real and an imaginary part. The most common way to analyze electrochemical impedance spectra is through the use of so-called equivalent circuits. An equivalent circuit is just a representation of the electrochemical system in terms of simple electric components such as resistors, capacitors and, in the presence of chemical reactions, even inductances [101].

The Randles circuit, depicted in Figure 5, is a common way to model a simple electrochemical system. The electrode–solution interface is represented by a capacitor, C_{dl}, representing the electrical double layer, in parallel with a second branch that accounts for a faradaic process. Such faradaic process consists of two components: a resistor, R_{ct}, which represents the charge transfer process, and a Warburg element Z_w that accounts for mass transport of the electroactive species from the solution bulk to the electrode surface. Next, in series with this parallel branch, we find a resistor, R_Ω, which represents the solution resistance between the WE and the RE. A similar circuit can be drawn for the auxiliary electrode, but since the WE processes are the limiting ones, the auxiliary electrode contribution is usually neglected. In addition, as will be discussed in the following sections, the feedback loop built driving the current at CE allows applying a known alternate current (AC) voltage between the RE and the WE, regardless of the impedance seen at the CE. This feedback loop has been indicated in the measurement setups of Figure 2 as a current-controlled generator, whose controlling quantity is the WE current, I_W.

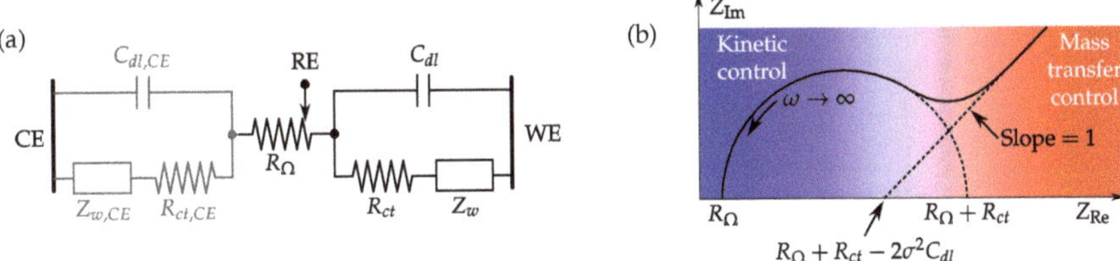

Figure 5. Randles equivalent circuit and Nyquist plot of electrochemical cell impedance seen between Counter electrode (CE) and working electrode (WE): (**a**) schematic; (**b**) characteristic on the real-imaginary plane.

In contrast with amperometric systems, where the electrode is driven to a condition far from equilibrium by the application of potential steps or potential sweeps, the EIS approach studies the system in a condition of quasi equilibrium. To this purpose, a small signal v_{ac} perturbation (usually <10 mV) is applied around a fixed V_{cell}, as depicted in Figure 2. The steady-state condition of the electrochemical cell allows for the linearisation of Equation (5) which leads to the expression of R_{ct}

$$R_{ct} = \frac{U_T}{n i_0}. \tag{6}$$

The Warburg impedance Z_w has a frequency (f) dependence law, and is commonly expressed as

$$Z_w(\omega) = \frac{\sigma}{\sqrt{\omega}}(1-\jmath); \quad \sigma = \frac{U_T}{n^2 FA\sqrt{2}}\left(\frac{1}{\sqrt{D_O}\,c_O^*} + \frac{1}{\sqrt{D_R}\,c_R^*}\right),\tag{7}$$

where $\omega = 2\pi f$, A is the area of the WE and D_O and D_R are respectively the diffusion coefficients of O and R. At high frequencies, $Z_w \to 0$, meaning that the diffusional process is too slow to influence the current. The characteristic $\omega^{-1/2}$ roll-off of Z_w prevents the use of simple circuital elements; however, RC approximations of Z_w are possible once a range of frequencies of interest is specified [102].

A typical approximated Nyquist plot (Imaginary vs. Real part) deriving from the Randles model is depicted in Figure 5b. It is worth noting that the series configuration of R_{ct} and Z_w reflects the current limitations of the faradaic process. When R_{ct} dominates, the system is kinetically limited, while, when Z_w dominates, the system is limited by the mass transfer phenomenon. The operating condition of the cell, i.e., the value of V_{cell}, determines which phenomenon is dominating.

To facilitate the discussion on the CMOS interfaces, we analyse here the spectral properties of the electro-chemical cell by introducing the following time constants:

$$\tau_k \doteq R_{ct}C_{dl}; \quad \tau_d \doteq (\sigma C_{dl})^2.\tag{8}$$

τ_k characterises the exponential relaxation when mass transport is negligible, while τ_d takes into account the diffusional process. Table 3 reports the expression for the WE impedance (Z) as a function of τ_k and τ_d, while Figure 6 shows the Bode and Nyquist plots for different values of τ_k/τ_d.

Table 3. Expressions of the real and imaginary parts of the electrochemical impedance of the Randles circuit in Figure 5.

	Expressions [91]	Expressions after Equation (8)
$\mathrm{Re}\{Z(\omega)\}$	$R_\Omega + \dfrac{R_{ct} + \sigma/\sqrt{\omega}}{(C_{dl}\sigma\sqrt{\omega}+1)^2 + \omega^2 C_{dl}^2(R_{ct}+\sigma/\sqrt{\omega})^2}$	$R_\Omega + \dfrac{1}{C_{dl}} \cdot \dfrac{\tau_k + \sqrt{\tau_d/\omega}}{(\sqrt{\tau_d\omega}+1)^2 + \omega^2(\tau_k + \sqrt{\tau_d/\omega})^2}$
$\mathrm{Im}\{Z(\omega)\}$	$\dfrac{\omega C_{dl}(R_{ct}+\sigma/\sqrt{\omega})^2 + \sigma/\sqrt{\omega}(C_{dl}\sigma\sqrt{\omega}+1)}{(C_{dl}\sigma\sqrt{\omega}+1)^2 + \omega^2 C_{dl}^2(R_{ct}+\sigma/\sqrt{\omega})^2}$	$\dfrac{1}{C_{dl}} \cdot \dfrac{\omega(\tau_k + \sqrt{\tau_d/\omega})^2 + \sqrt{\tau_d/\omega}(\sqrt{\tau_d\omega}+1)}{(\sqrt{\tau_d\omega}+1)^2 + \omega^2(\tau_k + \sqrt{\tau_d/\omega})^2}$

At this point, it is interesting to determine the frequency regions dominated by the two phenomena. It is evident in the Bode plot that the τ_k/τ_d ratio strongly influences the frontier between the Z_w-dominated and the $R_{ct}C_{dl}$-dominated regions. For τ_k in the order of τ_d, $|Z|$ presents a double slope characteristic: at low frequencies, $\omega \leq \tau_d^{-1}$, $|Z|$ presents a $-10\,\mathrm{dB/dec}$ roll-off, indicating the predominance of Z_w. For $\omega > \tau_d^{-1}$, the slope starts to follow a $-20\,\mathrm{dB/dec}$ slope.

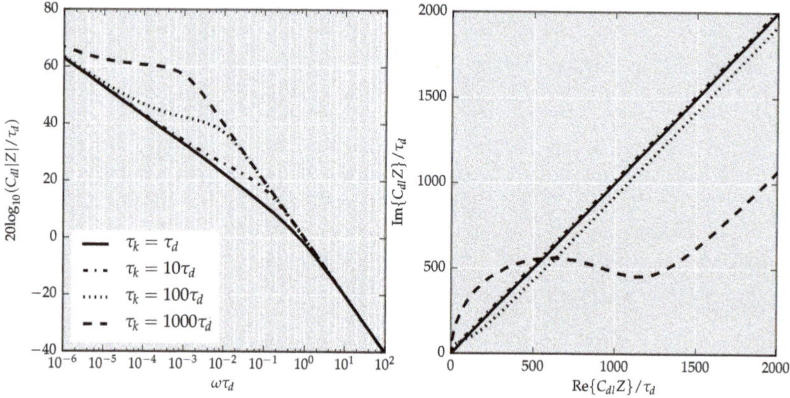

Figure 6. Bode (left) and Nyquist (right) plots for WE impedance Z. $R_\Omega = 0$ in all cases.

If the electrode kinetic is sluggish, i.e., $\tau_k \gg \tau_d$, a more complex behaviour is manifested, especially at low frequency. In this case, the τ_k^{-1} determines the pole location as in a standard RC circuit. At lower frequencies, the intersection with Z_w is still present, but it may occur far from the bandwidth of interest.

4. Advanced CMOS Interfaces for Electrochemical Sensors

4.1. Potentiometric Interfaces

Potentiometric sensing, based on the two-terminal measurement setup of Figure 2, relies on (1) to assess the concentration of a target analyte. The readout scheme of Figure 7, employed in the analog front-end (AFE) of the Melexis MLX90129 transponder (Melexis N.V., Ypres, Belgium), features two programmable gain amplifier (PGAs) to accommodate the weak potentiometric signal for the input fullscale of the ADC, after the removal of the constant baseline through a dedicated digital-to-analog converter (DAC). The system depicted in Figure 7, while providing a versatile architecture for a number of application cases, entails a certain degree of complexity. A simpler readout chain may be considered by removing the DAC and adopting a single amplification stage. This comes at the cost of increased resolution requirements for the ADC since now its fullscale is partially occupied by the constant term in Equation (1).

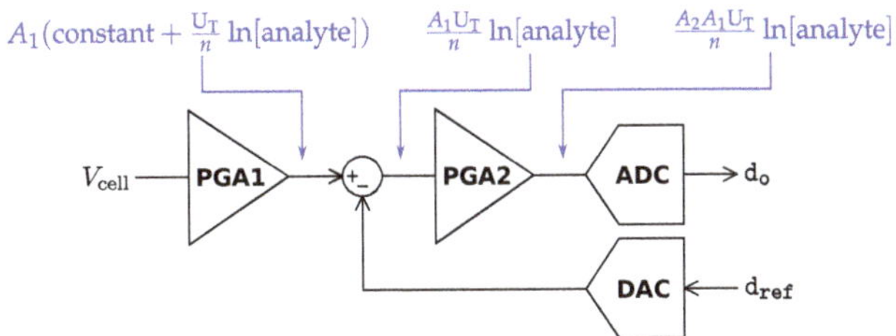

Figure 7. Generic potentiometric readout chain.

A detailed look at Table 1 reveals that, in some cases, specifications of the readout interface may require challenging resolutions in terms of input voltage levels. For instance, in the case of pH assessment in blood, a maximum variation <5 mV of the transduced signal is expected from Equation (1) considering the normal ranges reported in Table 1. Such weak signals are distributed in the sub-hertz frequency range where circuits' offset, offset thermal drift and flicker noise need to be cancelled in order to achieve and accurate and precise measurement.

Well-known circuital techniques for offset and low-frequency noise reduction in CMOS circuits are autozero (AZ), correlated double sampling (CDS) and chopper stabilization (CHS) [103]. Among them, CHS is immune to wideband noise fold-over into DC since it does not employ sampling. A part from intrinsic circuits' noise, environmental electrical interferences and spurious signals spread along circuits' bandwidth may be aliased in the base-band if they find a coupling path through the sampler. This could be easily the case of IoW devices, where large area sensors are coupled to their readout interfaces through printed conductive tracks onto a flexible substrate. Even in the ideal case of not having interferers, CHS is still to be preferred because it leaves a background noise equal to only one replica of the wideband thermal noise, while CDS and AZ, constrained by the bandwidth over sampling-rate ratio, would leave a certain multiple, usually ≥ 3, of the same background level.

Figure 8a shows a classical readout chain based on chopper stabilized instrumentation amplifier (In-amp). The input fully-differential modulator, usually implemented with four cross-coupled switches controlled by a two-phase, 50%-duty cycle digital clock, provides the up-conversion of the potentiometric signal around the chopper frequency f_{ch} and its odd multiples. The effect is equivalent to multiplying the signal by a dimensionless unitary square waveform, indicated with $m(t)$ in the figure. To fully benefit from CHS, f_{ch} should be high enough to move the modulated signal in a region where the flicker noise is not any more dominant over the wideband thermal noise.

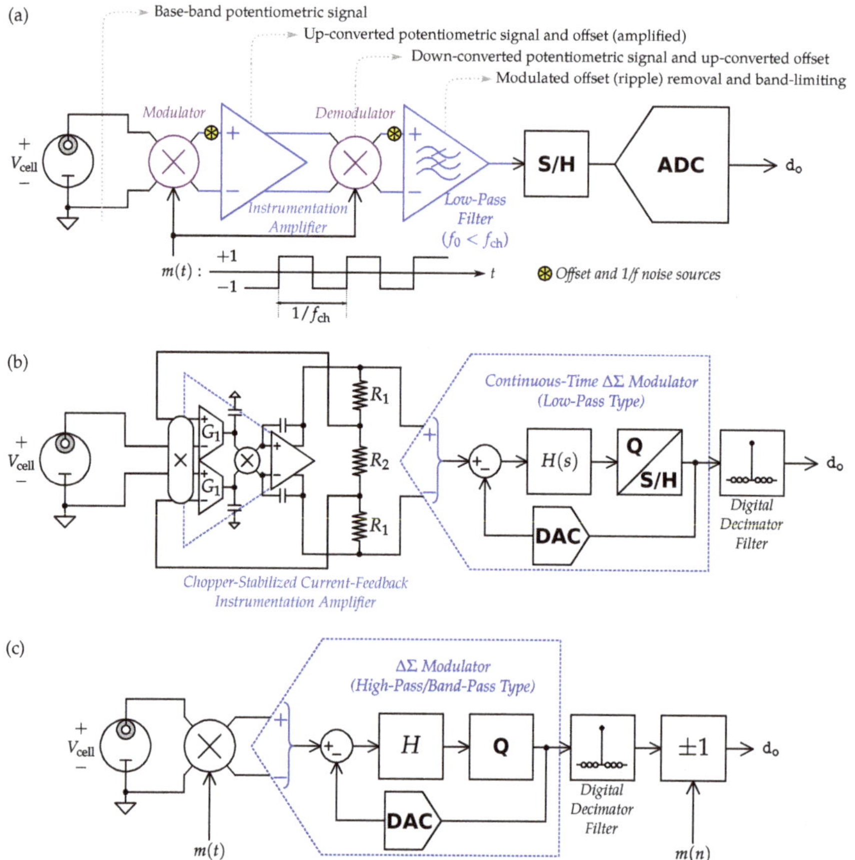

Figure 8. Readout chain variants for potentiometric signals: (**a**) classic chopper-stabilized redout chain; (**b**) improved by means of a modified current–feedback instrumentation amplifier; (**c**) based on early digitalization and system-level chopper stabilization.

However, several trade-offs influence the choice of f_{ch}: high chopper frequencies offer better flicker noise rejection and less area requirements for the In-amp and the subsequent low-pass filter [104,105], whereas low chopper frequencies reduce the residual offset and power requirements of the whole interface. In practice, chopper frequencies of few tens of kHz are used.

The In-amp processes, i.e., amplifies, its own offset and the upconverted potentiometric signal. The cascaded demodulator reverts the modulation by bringing back the potentiometric signal into the baseband and upconverting the amplifier offset. At this point, the offset which appears as a zero mean value square wave signal of $\pm A \cdot V_{OS}$ amplitude, where A is the In-amp precise gain and V_{OS} its input referred offset voltage.

A low-pass filter is needed in this case to interface the sample and hold circuit of the ADC for band-limiting and modulated component offset removal. At this stage, low-frequency noise and offset are added again to the potentiometric signal by the filter and the ADC, but their effect is less important since now the signal has been amplified. Nevertheless, the practical limitations apply since the the A is limited by the In-amp output fullscale and the filter input fullscale, which must not be exceeded by the sum of the amplified potentiometric signal and amplified modulated offset. This scenario is aggravated in the case of processing multiple channels, in a frequency multiplexing fashion [106].

A more advanced architecture, shown in Figure 8b, avoids the use of a low-pass filter by combining advanced chopper-stabilized In-amps, capable to auto-suppress the offset ripple, and the eventual use of continuous-time $\Delta\Sigma$-modulator which has intrinsic anti-aliasing characteristics [107]. Regarding advanced In-amps, a variety of solutions has been proposed in the last several years comprising selective band-pass In-amps [108] and current–feedback amplifier topologies featuring: ripple rejection feedback loops [109–112], use of switched-capacitors (SC) filters [113,114], ping-pong autozeroing [115], embedded low-pass filter [116,117] or digital calibration techniques [118,119]. Among all these techniques, the digitally-assisted solutions [118,119] proved to be the more efficient in terms of area and power. The ripple rejection technique proposed in [112] also promises good area efficiency, but silicon implementation has not yet been demonstrated.

A very important characteristic of CHS In-amps is their input impedance, which should be ideally infinite to avoid sensor loading. This is particularly true for ISEs, which are based on the Nernstian equilibrium of Equation (1) implying a zero net current flow at electrodes. Unfortunately, due to the combined effect of modulator activity and input parasitic capacitance of the In-amp, a net non-zero current is drawn from the sensor. Figure 9 helps understand this phenomenon. At each $m(t)$ transition, the input differential capacitance C_p undergoes a total voltage variation equal to $2V_{cell}$, accounting then for a charge $Q_p = 2C_p V_{cell}$ drawn from the sensor. When the opposite $m(t)$ transition occurs, approximately the same charge drawn from the sensor, an equivalent input resistance is estimated by

$$R_{in} = \frac{V_{cell}}{2Q_p f_{ch}} = \frac{1}{4 f_{ch} C_p}. \tag{9}$$

Considering $f_{ch} = 10\,\text{kHz}$ and an optimistic $C_p = 500\,\text{fF}$, Equation (9) estimates $R_{in} = 50\,\text{M}\Omega$, which would already imply a considerable current leakage flowing through the WE with the adverse effect of degrading the measurement precision.

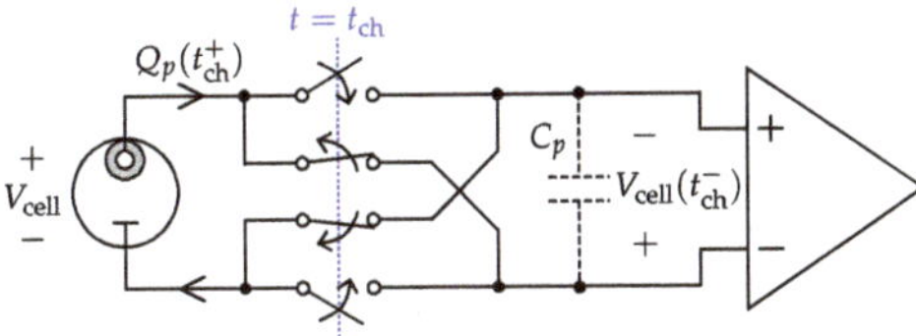

Figure 9. Potentiometric cell loading during transition of the input chopper modulator.

The chopped stabilized current–feedback amplifier proposed in [117] is particularly interesting because the swapping strategy of the input feedback ports of the In-amp demonstrated an input resistance boosted up to values exceeding $1\,\text{G}\Omega$ when the amplifier is operated at $f_{ch} = 20\,\text{kHz}$.

System level chopping, sketched in Figure 8c, is a quite mature concept [120,121]. It postulates the application of CHS around a generic ADC: the modulator operates on the input analog signal while the demodulator operates on the output converted digital signal. While input modulation can be achieved through the cross coupled switches, the demodulation operates directly in the digital representation of the signal by changing the sign of the the the data. System level chopping is very attractive since it enables early digitalization of the potentiometric signals greatly simplifying the readout chain and consequently enabling area and power reduction. A high-resolution readout can be achieved using a $\Delta\Sigma$ modulator, which needs to now deal with the up-converted potentiometric signal around at f_{ch} and its odd multiples. Consequently, a band-pass/high-pass modulator type is needed instead of the traditional low-pass type. A general rule to derive pass-band modulators from their low-pass model, is to substitute the integrator blocks with their resonator counter parts.

When an SC solution is considered, the resonator blocks are efficiently realized through the pseudo-2-path SC resonators [107] (p. 156). The same block applied to a CHS $\Delta\Sigma$ converter has been recently analyzed to highlight its functional characteristics for various $f_{ch}/f_{sampling}$ ratios [122].

An interesting silicon implementation of the same idea is provided in [123] consuming only 144 μW and reducing the input offset to 403 μV. It is worth noting that SC implementations of the concept in Figure 8c, where no anti-aliasing filter is present, may introduce significant degradation of the converter resolution due to noise fold-over.

Alternatively to ΔΣ-modulators, Successive Approximation Register (SAR) ADCs can be employed for digitalization. Ref. [124] presents a potentiometric AFE based on an SAR ADC constrained to a power budget for the sensor AFE of <1.2 pW and capable of 8.3 ENOB of resolution. Such a remarkable ultra-low power feature is achieved thanks to a reference-free charge-sharing SAR architecture which avoids the use of power hungry voltage reference buffers [125]. In addition, since an extremely slow sample rate is used (10 S/s), a combination of two different and clock generation strategies is used. The sampling is controlled by an ultra-low-power capacitve discharging oscillator running at 10 Hz [126], while the conversion-step clock is generated by an asycronous unit running at 1 MHz for the 10 controlling slices of the conversion process [127]. This strategy allows for greatly relaxing leakage problems of the charge sharing architecture when operated at extremely low switching frequency.

Despite the great amount of techniques and circuital possibilities already known, research efforts are still needed to find optimal potentiometric interface solutions for the IoW.

4.2. Amperometric Interfaces

A potentiostatic interface performs three functions: (i) it sets the cell operating voltage V_{cell}; (ii) it reads out the redox current at WE, I_W; and (iii) it provides analog-to-digital (A/D) conversion. Figure 10 presents general topologies for potentiostats that perform the amperometric readout relying on current to voltage (I_W-to-V_A) conversion through a resistor R prior its A/D conversion [128–131]. The most common configuration is shown in Figure 10a. Here, A_1 and A_2 set V_{RE} and V_{WE}, respectively, through local feedback loops. While A_1 feedback is built around the sensor to set the RE potential, A_2-R realize the well-known trans-impedance amplifier (TIA) configuration while setting at the same time the WE potential. If A_1 and A_2 are ideal operational amplifiers, OpAmps, straightforward circuit analysis reveals

$$V_{\text{cell}} \doteq V_{RE} - V_{WE} \approx V'_{RE} - V'_{WE}; \quad V_A = RI_W, \tag{10}$$

where the high-gain operational amplifiers ($A_1 \gg 1$ and $A_2 \gg 1$) ensure that $V_{RE} \approx V'_{RE}$ and $V_{WE} \approx V'_{WE}$. Equation (10) shows that V_{cell} can be set by independently varying V'_{RE} or V'_{WE}. The same equations apply to the circuit in Figure 10b. However, in this configuration, the sensor in included in both A_1 and A_2 loops. More importantly, the feedback current of A_2 in Figure 10b flows through both R and the sensor. This way, the sensor is included in the readout loop, giving rise to very compact topologies that will be commented on at the end of this Section.

The potentiostats of Figure 10a–c provide a very wide swing of V_{cell}, ideally doubling the effective supply voltage, provided that the DAC outputs, V'_{RE} and V'_{WE}, can swing across the full span of the voltage supply rails. It is also clear that the common-mode input range of the amplifiers in schemes (a) and (b) must also be rail-to-rail since one of the inputs is connected to either V'_{RE} or V'_{WE}. Nevertheless, voltage limitations on the V_{cell} swing are evident once the output of A_2 (V_{out,A_2}) is considered. Without loss of generality, we assume that a single supply voltage (rails gnd, V_{DD}) is available. For both schemes (a) and (b),

$$V_{out,A_2} = V_{WE} - RI_W. \tag{11}$$

Let us first consider the case of $V_{\text{cell}} > 0$ and $I_W > 0$: in this situation, V_{out,A_2} tends to lower voltage values, in order to avoid saturation of A_2,

$$V_{out,A_2} > 0 : V_{WE} - RI_W = V_{RE} - V_{\text{cell}} - RI_W > 0 \Rightarrow V_{RE} > V_{\text{cell}} + \underbrace{RI_W}_{V_A}. \tag{12}$$

By conveniently setting $V_{RE} \to V_{DD}$, the maximum room for voltage allocation between V_{cell} and V_A can be obtained. Conversely, for the case of $V_{cell} < 0$ and $I_W < 0$, V_{out,A_2} tends to saturate towards V_{DD}:

$$V_{out,A_2} < V_{DD} : V_{WE} + R|I_W| = V_{RE} + |V_{cell}| + R|I_W| < V_{DD} \Rightarrow V_{DD} - V_{RE} > |V_{cell}| + \underbrace{|RI_W|}_{|V_A|}. \quad (13)$$

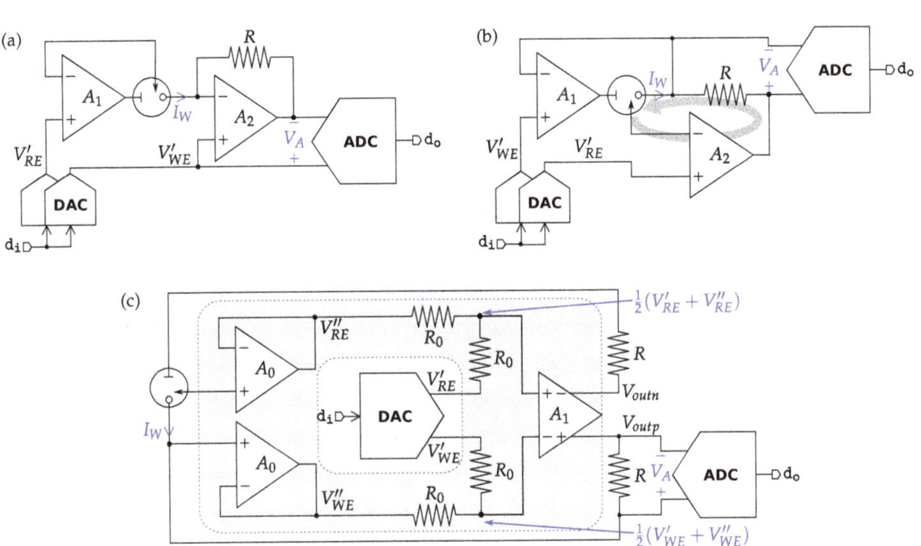

Figure 10. Potentiostatic interfaces based on continuous time *I*-to-*V* conversion (**a**) conventional solution; (**b**) sensor-in-the-loop solution; (**c**) fully-differential solution.

Thus, voltage range is maximized by setting $V_{RE} \to 0$. Equations (12) and (13) clearly show the trade-offs to be considered, between the potentiostatic range (V_{cell}) and the trans-resistive gain, R, that can be achieved. This trade-off is particularly severe when a low-voltage design is targeted.

Fully differential topologies such as those in Figures 10c have also been proposed [131–133]. The solution in [133] replaces all the gray area in Figures 10c with a fully-differential difference amplifier (FDDA). Analytically, considering the Randles model of Figure 5 and considering that no redox process is present at CE,

$$V_{CE} = V_{outn} - RI_W \approx V_{RE} + R_\Omega I_W; \quad V_{WE} = V_{outp} + RI_W. \quad (14)$$

Now, expressing each differential output voltages V_{outn}, V_{outp}, it will be easily found that

$$V_{cell} = \frac{A_1}{1 + A_1}(V'_{RE} - V'_{WE}) - \frac{(2R + R_\Omega)I_W}{1 + A_1} \underset{(A_1 \gg 1; R \gg R_\Omega)}{\approx} V'_{RE} - V'_{WE} - \frac{2V_A}{A_1}. \quad (15)$$

Equation (15) reveals that both R_Ω and R play a detrimental role on the precision of potentiostat operation, and can only be compensated by large values of A_1. While $R_\Omega \ll R$, and it can thus be neglected in many practical cases, the issue is still present due to the fact that R is usually as large to accommodate V_A to the input full scale of the ADC. Therefore, the error on V_{cell} can be approximated to $2V_A / A_1$. Moreover, scheme (c) is still subject to the trade-offs of Equations (12) and (13).

The same equations can be reinterpreted to find the specifications for the input common mode swing of the ADCs. All the schemes shown in Figure 10 imply demanding requirements for input common mode range as it needs to span from 0 to V_{DD} to provide the maximum voltage room for

V_{out,A_2}. The use of an instrumentation amplifier, as interstage between the *I*-to-*V* stage and the ADC, may provide a suitable common mode matching and further gain. On one hand, distributing the gain among various stages may result in being beneficial for medium/high bandwidth application [134], but this could rarely be the case for IoW interfaces. On the other hand, R also defines the noise floor asymptote of the amperometric readout, which finally determines the amperometric limit of detection (LOD) of the potentiostat:

$$S_{i_n} = \frac{4KT}{R}; \quad \text{LOD}_{\text{ideal}} = 3\sqrt{S_{i_n} \times \text{bandwidth}}. \tag{16}$$

In practice, extremely large values of R (>10 MΩ) are avoided since the integration of such high resistances is too expensive in terms of silicon area. The use of off-chip resistors is feasible, but it comes at the cost of extra pads. By looking at Equation (5), it can be seen that the amperometric sensitivity can be incremented by enlarging the sensing area of the WE, which is already an off-chip component of the envisioned IoW system, thus avoiding the use of extremely large R and relaxing the LOD requirements.

Amplifiers' noise is also present and their effect in the amperometric readout has been studied in [135] for simplified Randles circuit ($Z_w = 0$). From analysis in Section 3.3, this is the case of a sluggish process, i.e., when the electrode kinetics are dominated by the charge transfer mechanism. More generally, a beneficial interplay between the sensor impedance and the $1/f$ noise components exists at low frequencies and will be addressed here. In Figure 11, referred-to-input noise of A_1, A_2 has been introduced. Small-signal equations can be written:

$$\begin{cases} i_n = (v_{CE} - v_{WE})/Z \\ v_{CE} = R_\Omega i_n + v_{RE} \end{cases} \longrightarrow \quad i_n = \frac{v_{CE} - v_{WE}}{Z - R_\Omega} : \quad S_{i_n} = \frac{S_{v_1} + S_{v_2}}{|Z - R_\Omega|^2}. \tag{17}$$

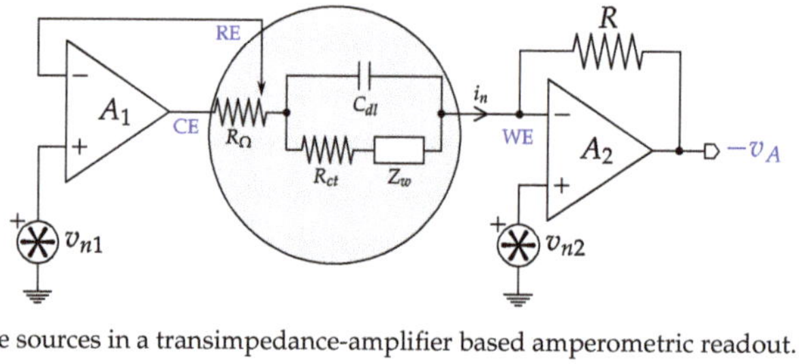

Figure 11. Noise sources in a transimpedance-amplifier based amperometric readout. Configuration (**a**) of Figure 10 is considered here.

Since v_{n1}, v_{n2} are originated from different amplifiers, they should be considered as uncorrelated noise sources. At very low frequencies, $f < f_c$, where $Z \approx Z_w$, we can use the flicker noise expressions for S_v:

$$S_v|_{f<f_c} = k_A \frac{K_F}{WL} \frac{1}{|f|} \longrightarrow \quad S_{i_n}|_{f<f_c} = \frac{K_F}{4\sigma^2}\left[\frac{k_{A_1}}{(WL)_1} + \frac{k_{A_2}}{(WL)_2}\right], \tag{18}$$

K_F being the flicker coefficient of the MOS transistors for the given technology; W, L the geometric characteristics of the amplifiers' input devices and k_{A_1}, k_{A_2} the respective excess noise factors of A_1 and A_2 taking into account the circuital implementation. Equation (18) shows clearly that the Warburg impedance provides built-in flicker noise stabilization of the electronic interface. It is important to notice that σ depends on the analyte concentration (both in its reduced and oxidation state), so the total noise power is expected to be signal dependent.

Another noise contribution must be accounted for when considering the readout point after the TIA stage. Any fluctuation caused by v_{n2} propagates to the output of the TIA regardless of Z, so Equation (17) underestimates the total noise contribution. A more complete expression can be found by direct inspection of the circuit of Figure 11:

$$\begin{cases} v_{A,\text{noise}} = \left(1 + \dfrac{R}{Z - R_\Omega}\right) v_{n2} - \dfrac{R v_{n1}}{Z - R_\Omega} \\ i_{n,\text{rti}} = \dfrac{v_{A,\text{noise}}}{R} \end{cases} \longrightarrow S_{i_{n,\text{rti}}} = \dfrac{S_{v_1} + S_{v_2}}{|Z - R_\Omega|^2} + \left(1 + \dfrac{2R}{\text{Re}\{Z - R_\Omega\}}\right) \dfrac{S_{v_2}}{R^2}. \tag{19}$$

A multifaceted spectral scenario must be considered for the calculation of the LOD once the amplifiers' noise contribution is taken into account:

$$\text{LOD} = 3 \times \left[\int_{T_{obs}^{-1}}^{\text{bandwidth}} S_{i_{n,\text{rti}}}(f)\, df \right]^{\frac{1}{2}} + \text{LOD}_{\text{ideal}}. \tag{20}$$

Here, T_{obs} is the observation time of the measurement. Practical limits apply to the validity of Equation (18) for very large T_{obs}, and the diffusion layer is so extended that other phenomena detrimentally influence the measurement. Among such phenomena, the most relevant are convection and sample evaporation. Due to the high flicker noise levels present in CMOS circuits, high accuracies require the use of dynamic techniques for $1/f$ noise reduction, such as CHS, AZ or CDS.

Another well known continuous time amperometric readout technique is based on current mirror amplifiers [129,130,134,136–139], such as those shown in Figure 12. A closer look at the configurations in Figure 10 reveals that the potentiostatic control amplifier A_1 provides the current to the sensor, so it can be copied and eventually amplified by simple current mirror configurations.

In Figure 12a, the mirror is embedded in the amplifier, and current is precisely copied with the help of an auxiliary amplifier A_2 and a cascode transistor (M_C) [129]. The sensor current is amplified thanks to the multiplier factor m of the current mirror. It is worth noting here that the sensor readout current biases M_1, which also provides an additional gain stage in the amplification loop. Thus, when the current is extremely low, as in case of calibration with a blank sample, the control loop may be easily pushed to instability. A minimum bias current I_0 can be added at the input of the mirror to circumvent this problem.

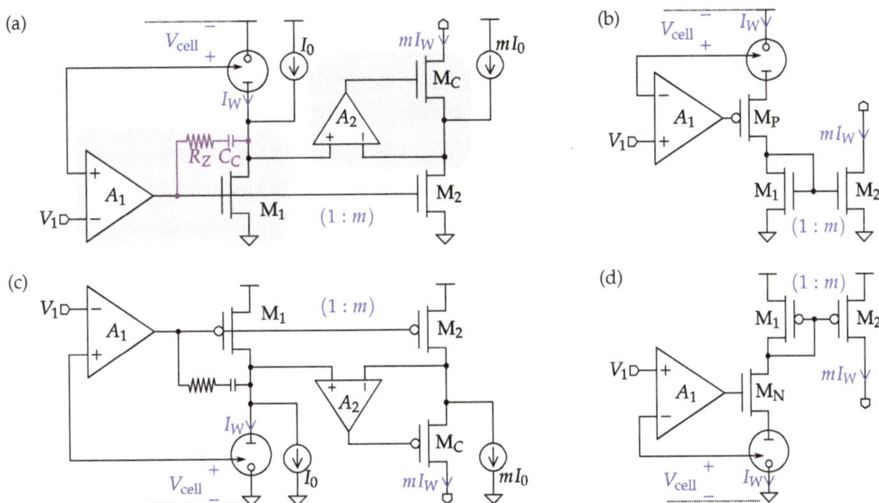

Figure 12. Potentiostats based on current mirror amplifiers. (**a**) n-type mirror embedded in the amplifier; (**b**) n-type mirror with current buffer; (**c**) p-type mirror embedded in the amplifier; (**d**) p-type mirror with current buffer.

Stability can be further enhanced by introducing an LHP zero in a feed-forward signal path around M_1 [136]. Schemes (a) and (c) shows a simple zero-nulling through a resistor R_Z; however, other optimum compensation schemes may be employed depending on the actual circuital implementation of A_1 [140].

Configuration in Figure 12b employs a p-type device M_P acting as a common drain stage rather than a common source stage, thus providing a current buffer between the control loop and the current mirror to improve stability. The simpler topology of Figure 12b comes at the cost of reduced headroom for V_{cell} due to the V_{GS} drop of M_P. While both topologies are attractive for high-bandwidth applications and for the low-count of devices, noise performance can be of concern since $1/f$ components of M_1-M_2 are not filtered through the impedance of the sensor and they sum up directly to I_W. On the other hand, their potentiostatic capabilities are also limited by the fact that the WE is fixed to the positive supply voltage, so only negative V_{cell} can be applied. The complementary topologies of Figure 12c,d, i.e., interchanging n-type and p-type devices, are also possible, which would allow the application of positive values of V_{cell} only.

High-sensitive readout can be achieved by replacing the resistive feedback element with a capacitor, following the SC paradigm. Intuitively, R in Equation (19) is now substituted by $1/(j\omega C)$, so S_{v2} noise contribution is greatly attenuated in the low frequency range. In such case, we can express V_A between the integration frame delimited by T_{i-1} and T_i as

$$V_A(T_{i-1}, T_i) = \frac{1}{C} \int_{T_{i-1}}^{T_i} I_W(t)\, dt = \frac{T_i - T_{i-1}}{C} \cdot \overline{I_W}, \tag{21}$$

which reveals that the readout value V_A holds the averaged I_W current within the integration frame.

The use of a capacitive element allows the I-to-V interface to reduce area occupation, and to benefit from better linearity and less PVT variations with respect to resistive elements. However, a sustained current would soon drive to saturation the integrator so some sort of reset strategy is needed [141–147].

A classic SC reset mechanism that also performs CDS is shown in Figure 13 [142,143,147]. The scheme in (a) features a PGA built around A_3, whose (inverting) gain is set by C_1/C_2. Here, during the reset phase φ_1, v_{n2} and v_{n3} are sampled respectively onto C_1 and C_2.

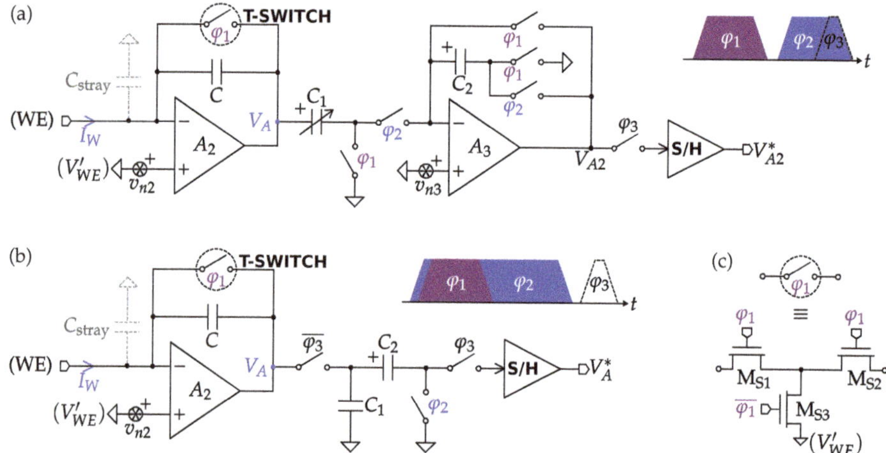

Figure 13. Switched-capacitors amperometric readout schemes including flicker noise cancellation by correlated-double sampling. (**a**) conventional scheme; (**b**) without the programmable-gain amplifier; (**c**) T-switch schematic.

In the integration phase φ_2, the signal follows the integration and the amplification stages together with the current values v_{n2} and v_{n3}, which now contributes with the opposite sign. Noise sample differentiation of the CDS techniques is thus readily implemented. At the end of ϕ_2, the sample and hold (S/H) phase ϕ_3 delivers the processed output V_{A2}^* to the following A/D stage. Scheme (b) follows the same philosophy as (a), without the PGA stage [135]. A different arrangement of the phases allows for signal integration and noise differentiation with only one active element.

While SC readout chains offer the switching frequency as a valuable degree of freedom to the designer, especially to avoid integration of large R on chip, the achievable resolution is affected by several issues, which are specific to SC circuits, such as: noise folding due to the sampling process at C_1 and C_2 (kT/C noise), amplifier noise, charge injection, clock feed-through and reset switch leakage at the integrator stage.

Noise folding can be cut down by setting relatively large values of C_1 and C_2. As detailed in [148] and references therein, amplifier noise can be embodied into kT/C noise as a form of excess noise that can be limited by a proper design. In practice, values of C in the order of $\approx$100 fF are sufficient to make kT/C noise contributions smaller than ADC resolution. Charge injection and clock feed-through can be reduced through fully-differential architectures, dummy switches [103] and advanced layout techniques [149]. Leakage of the reset switch can be minimized by using the t-switch configuration shown in Figure 13c [150]. In this case, leakage currents of M_{S1} are nulled by equalizing the source/drain voltage through the virtual ground on the left side and the shunt connection through M_{S3} on the right side.

So far, the A/D conversion step in the signal processing flow has been considered as a separate function carried out by a dedicated block, namely the Analog-to-Digital Converter (ADC). This classic approach is followed in [146], where an SAR ADC has been coupled to an SC-integrator based readout chain. SAR ADCs represent a popular solution in the medium-low resolution range thanks to their excellent power efficiency. On the other hand, ADC based on $\Delta\Sigma$ modulators is also an attractive choice for high resolution low-bandwidth applications like the IoW. While $\Delta\Sigma$ ADCs needs a digital decimation filter for signal reconstruction from the $\Delta\Sigma$ bit stream produced by the modulator, they employ the oversampling ratio (OSR) to relax both thermal noise and matching constraints.

At the same time, the need for early digitalization inspired a number of compact readout architectures providing fully or semi digital outputs. Figure 14 illustrates the main difference in potentiostatic functioning principles between the classic and the compact approach. The compact approach skips the I-to-V conversion step in favour of direct I_W digitalization. In some cases that will be discussed next, the same digital output is fed back to the potentiostat in order to set the V_{cell}. In the remainder of this section, potentiostats will be classified as:

1. Digital potentiostat by the means of Dual-Slope A/D conversion [141,150];
2. I-to-f potentiostats: the readout current is converted to a square-wave signal whose frequency is proportional to the current amplitude [136–139,151];
3. I-to-PDM (pulse density modulation) potentiostats: the readout current modulates the density of a pulse repetition in time [128,152–154];
4. I-to-$\Delta\Sigma$M potentiostats: the readout current is converted into a stream of $\Delta\Sigma$ modulated signals [150,155–161].

Smart potentiostats falling into the first category employ a capacitive TIA (CTIA) stage whose input is fed alternatively by the I_W current and a reference discharge current I_{ref}. Figure 15 shows the simple circuital arrangement and the timing of the dual-slope configuration. During the integrating phase, I_W charges C either to cause a positive or negative ramp of V_C. The total integrated charge will be $T_{int}I_W$. At the end of this phase, the comparator decides the polarity of I_W in order to determine the I_{ref} to be of source or sink type during the controlled discharge phase. A digital counter, embedded in the digital control block, starts to run until the comparator reverts to its initial state; at this point, the I_{ref} stops integrating onto C, thus the charge variation during this phase will be $T_{dis}I_{ref}$ in magnitude, and its sign will be opposite with respect to the charge accumulated during integration. Since the comparator triggers on the same voltage level, the net charge change over C in the current conversion cycle is zero ($T_{\text{int}}I_W = T_{\text{dis}}I_{\text{ref}}$), and the counter output will be represented by the digital value of:

$$d_{\text{out}} = f_{\text{ck}}T_{\text{dis}} = f_{\text{ck}}T_{\text{int}}\frac{I_W}{I_{\text{ref}}}. \tag{22}$$

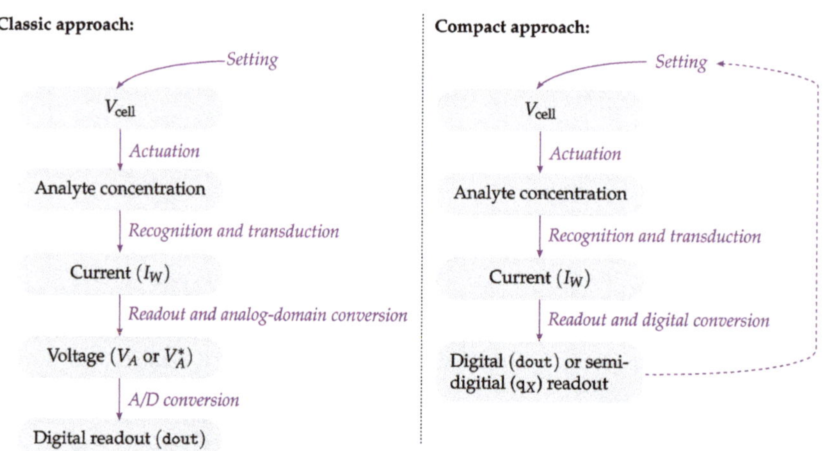

Figure 14. Classic and compact potentiostat approaches.

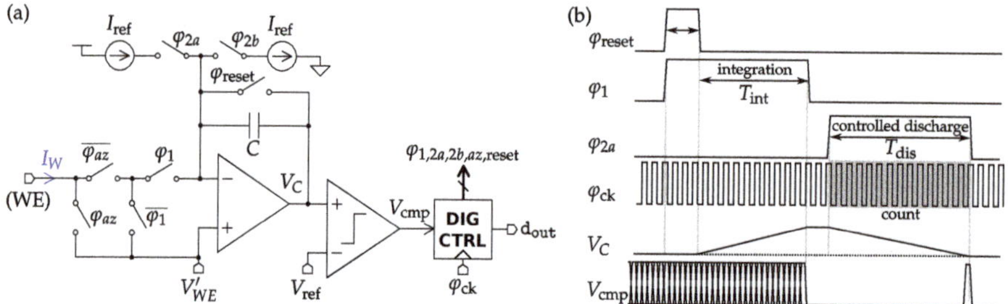

Figure 15. Potentiostats with embedded dual slope A/D conversion. (**a**) schematic; (**b**) time diagram.

Potentiostats based on current to frequency conversion (I-to-f) operate in a similar manner to the dual slope potentiostats, controlling the charging and the discharging of a capacitor element [128,136]. However, instead of directly converting the discharge time into a digital count, the period of a square waveform is continuously modulated which can be later converted into a digital value by time discretization.

The circuit in Figure 16a performs the I-to-f conversion through a simple relaxation oscillator constraining V_C between $V_{\text{ref,H}}$ and $V_{\text{ref,L}}$. Here, I_W is injected into the the I-to-f modulator through a current amplifier potentiostat such as those of Figure 12. Similarly, the circuit in Figure 16b integrates I_W alternatively on two nominally identical capacitors C_1 and C_2 [138,139]. Here, the use of a single comparator triggering at a fixed voltage makes the modulated square wave insensitive to comparator offset (and to its temperature drift). Furthermore, temperature effects on the modulated signal can be cancelled out by performing a double reading, first measuring the PTAT modulated signal, and then adding the sensed current contribution:

$$f_{\text{ref}} = \frac{I_{\text{PTAT}}}{(C_1 + C_2)V_{th}}; \quad f_{\text{meas}} = \frac{I_{\text{PTAT}} + I_W}{(C_1 + C_2)V_{th}}. \tag{23}$$

The solution in Figure 16c proposes an I-to-V conversion through a starved-inverter based ring oscillator [151]. Interestingly, this interface provides a full potentiostatic control at WE through the action of a current conveyor type-II (CCII+) circuit. The voltage set at the high impedance input Y is replicated onto X providing a virtual ground; moreover, the CCII+ senses the current at the low-impedance input X and replicates it at the Z terminal. Since Z is connected to the gate of an array of transistors (M_{R1}–M_{RN}), no current is allowed to flow; consequently, the control loop imposes $I_X = 0$ and all I_W is conveyed at one of the M_R transistors corresponding to the starved inverter activated by the travelling wave. Only one CCII+ is needed for all the ring stages, since the regulation control of each M3 is naturally rolling along the ring in synchronism with the steering of I_W. This particular sequential control gives the name rolling regulated current-controlled ring oscillator to the circuit of Figure 16c. Differently from standard unregulated current-starved ring oscillators, voltage threshold to be reached by the travelling edge here is set precisely to be V'_{WE}. Thus, the transit time of for each ring slice is determined by:

$$T_{\text{slice}} = \frac{CV'_{WE}}{I_W},$$
(24)

and the oscillation frequency, which is a submultiple of $1/T_{\text{slice}}$, will be linearly proportional to the sensed current I_W. The rolling regulation allows very linear characteristics of the I-to-f conversion without the need for any digital calibration or post-compensation technique. The digital readout is performed by combining the full rounds of the travelling edge detected by the course counter, and the intermediate position of the edge inside the loop extracted by the fine encoder. This technique enables extended resolution for the same frequency range at the cost of chip area, without increasing power consumption.

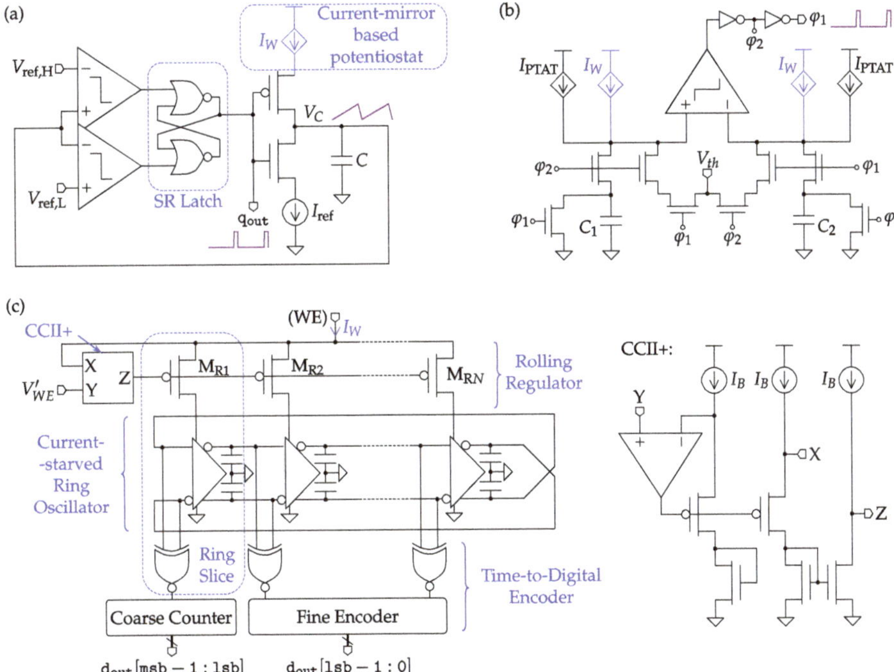

Figure 16. Potentiostats with embedded current to frequency conversion (I-to-f type): (**a**) based on conventional relaxation oscillator; (**b**) based on single-comparator relaxation oscillator; (**c**) based on current-starved ring oscillator.

A natural extension of of I-to-f techniques are those involving a fixed-width pulse whose repetition period is controlled by the modulating current. This way, the repetition frequency of the pulse, which is also the pulse density in time, encodes the I_W information. I-to-PDM potentiostats can have a minimalistic circuital implementation as shown in Figure 17. The scheme in (a) derives from that in Figure 12b, where the current mirror is replaced by a simple integrate-and-fire (IaF) circuit composed by a shunt capacitor and a comparator [152]. Starting from a reset condition, V_C increases its value at the constant rate of I_W/C; when V_C reaches the threshold value set by V_{DAC}, an event is fired, whose duration T_{pulse} may be controlled by the digital control circuit. At the event onset, the capacitor is reset and T_{pulse} should be long enough to ensure the complete reset of the capacitor. The firing rate f_{ev} is thus determined by:

$$f_{ev} = \frac{1}{\dfrac{CV_{DAC}}{I_W} + T_{pulse}} \underset{(T_{pulse} \to 0)}{\approx} \frac{I_W}{CV_{DAC}}. \tag{25}$$

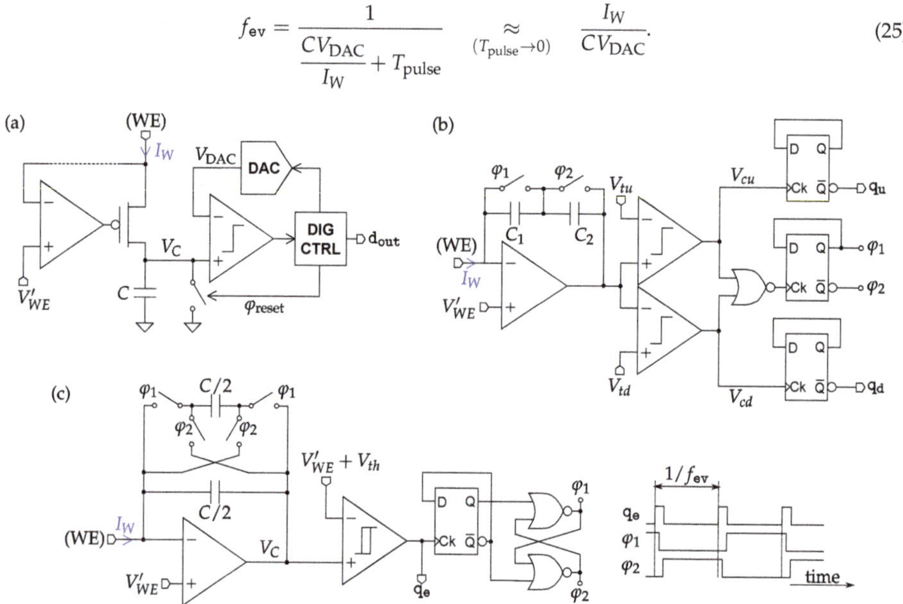

Figure 17. Potentiostats with embedded current to pulse density modulation (I-to-pulse-density-modulation type). (**a**) based on current buffer; (**b**) based on transimpedance amplifier with double integrating capacitor; (**c**) based transimpedance amplifier with soft-reset mechanism.

Equation (25) clearly shows that a non-zero T_{pulse} implies a nonlinear $f_{ev}(I_W)$ characteristic. Thus, a very small value of T_{pulse} is desirable. On the other hand, if T_{pulse} is not long enough to reset V_C completely, the next pulse would occur after a delay that depends not only on I_W but also on the residual charge; since the residue depends on the discharging characteristic of the nonlinear MOS switch, it will not be linear with respect to its associated V_C sample, finally creating a strongly nonlinear I-to-PDM conversion characteristic.

In order to maintain a sufficiently linear response for a given T_{pulse}, V_{DAC} can be coarsely adjusted following a discrete set of I_W ranges. This way, the integration time can be sufficiently larger than T_{pulse} for a large range of I_W values.

The circuit of Figure 17b avoids the problem related to a non-zero T_{pulse} by alternating the integration and reset phase on two nominally identical capacitors C_1 and C_2 [153]. The modulator also features a double comparator to detect separately positive and negative I_W.

The modulator proposed in Figure 17c is also insensitive to T_{pulse}, without doubling the integration capacitor [154]. Instead, C is split in two, and reset is performed by flipping one of the halves and thus neutralizing the total integrated charge. More importantly, schemes (b) and (c) both perform charge neutralization without the intervention of the amplifier, greatly relaxing its speed requirements.

The last class of smart potentiostats involves those performing current to $\Delta\Sigma$-modulated bit stream. In such circuits, the signal is processed through integration and comparison to a reference voltage, as in the previous solutions. However, the comparison takes place at a constant pace; therefore, the output signal is fully digital since both its amplitude and its duration are of a discrete nature. From the circuital point of view, this avoids the use of continuous-time comparators in favour of latched-type comparators which enable important DC power reduction [162].

In the circuit of Figure 18a, the $\Delta\Sigma$ loop is built around the CTIA stage in a similar fashion to the dual-slope potentiostat of Figure 15a. However, phases' control is different here since, at the same time I_W is integrated, the sinking or sourcing I_{ref} is also contributing to the error charge accumulated onto the integrator. At each quantization cycle, the error sign is extracted and the loop counteracts in order to cancel it. On a large time scale, the average error is nulled, thus making the averaged quantized stream q_{dsm} also a digital representation of the input current [155].

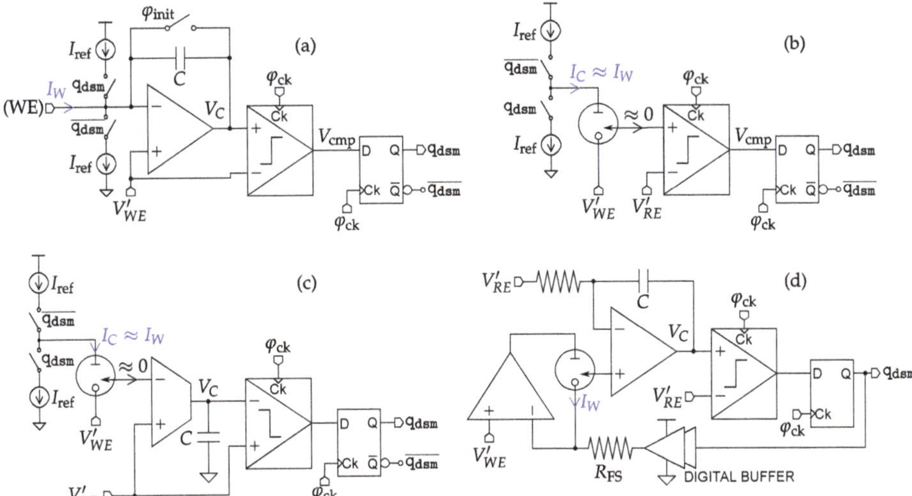

Figure 18. Potentiostats with embedded $\Delta\Sigma$ modulation (I-to-$\Delta\Sigma$M type). (**a**) based on transimpedance amplifier; (**b**) sensor-in-the-loop approach, first-order modulator; (**c**,**d**) sensor-in-the-loop approach, second-order modulators.

The circuit of Figure 18b also implements a $\Delta\Sigma$ loop, but this time the sensor itself, with its intrinsic low-pass quasi-integrator spectral characteristic is used as the loop shaper (see Z expressions in Table 3).

The voltage error signal is extracted at RE allowing the current fed through the WE to compensate for it. This error is a small ripple around the V'_{RE} set point chosen for the RE. Since the input impedance of the comparator is much smaller with respect to the intrinsic C_{dl} of the cell, the alternating current provoked by the RE ripple is negligible and the condition of nearly zero current at RE is ensured. It is worth noting that scheme (b) employs the sensor-in-the-loop strategy mentioned at the beginning of this section. This way, a great compactness of the interface is obtained, which also allowed the fusion of the potentiostatic control and the amperometric readout in a single control loop.

Unfortunately, the minimalistic $\Delta\Sigma$ modulator of Figure 18b suffers from poor potentiostatic linearity [158] due to signal dependent gain of the single-bit quantizer [107]. Topologies (c) [159] and (d) [160] in the same figure solve this problem by introducing an integrating stage before the quantizer that ensures a much more precise potentiostatic control of the interface. Moreover, since now a mixed electrochemical and pure electronic shaper is employed, the modulator order is increased providing higher signal-to-noise ratios for the same readout bandwidth. It is important here to emphasize that modulators in (b–d) implement the dashed feedback loop depicted previously in Figure 14, where the the digital readout is reused internally in the interface to set the V_{cell} potentiostatic voltage.

The vast majority of the state-of-the-art works reviewed here employ uniform and continuously operating sampling. In some cases, low levels of signal for prolonged periods of times are followed by sparse burst-like activity. In another scenario, a more dense sampling is desired after the signal of interest triggers a certain threshold. In both cases, smart adaptive sampling can provide a more efficient energy management by lowering the data rate during idle periods. Typically, the very low bandwidth requirements of the many kinds of physiological signals allow for aggressive duty cycling, with large scheduled off-state periods. Thus, in energy-constrained scenarios such as the IoW, adaptive sampling rate and duty cycling will constitute the basis of the next-generation interfaces.

4.3. Interfaces for Sensing FETs

The idea of using a field-effect transistor (FET) as an active pH sensor was first published more than 45 years ago [163,164]. However, a major milestone in the history of the ion-selective FET (ISFET) arrived almost three decades later with the demonstration of its fabrication through unmolested standard CMOS technologies [165]. As a result, today's ISFETs benefit from the large-scale integration and low-cost manufacturing advantages of the *More Moore* and *More than Moore* technology roadmaps [166]. Recently, authors in [167] fabricated an ISFET on a flexible substrate demonstrating its employability as a wearable sweat pH sensor for healthcare.

Basically, an ISFET device is a regular MOS transistor with its floating gate isolated from the rest of the CMOS circuit but coupled to an external RE (typ. Ag/AgCl) in contact with the electrolyte solution through an insulating membrane (typ. CMOS standard passivation Si_3N_4/SiO_2) as the sensing area. Since no DC current is driven to the gate, the ISFET can be understood as a kind of potentiometric chemical sensor, although in most literature references its pH sensing capability is explained as a change in the threshold voltage of the corresponding floating-gate MOS device.

The equivalent electrical model of the ISFET is depicted in Figure 19. The electrochemical cell (V_{chem}) consists of the RE built-in potential (V_{ref}) together with the double-layer capacitance formed by the Gouy–Chapman layer (C_{gouy}) and the Helmholtz plane (C_{helm}) [168]. Indeed, the latter defines the pH-driven voltage drop (V_{ph}) at the MOSFET gate. The rest of the ISFET model of Figure 19b includes the passivation capacitance (C_{pass}) and the inner potential drop (V_{tc}) introduced by the charges trapped in the insulator layers during the MOSFET integration process.

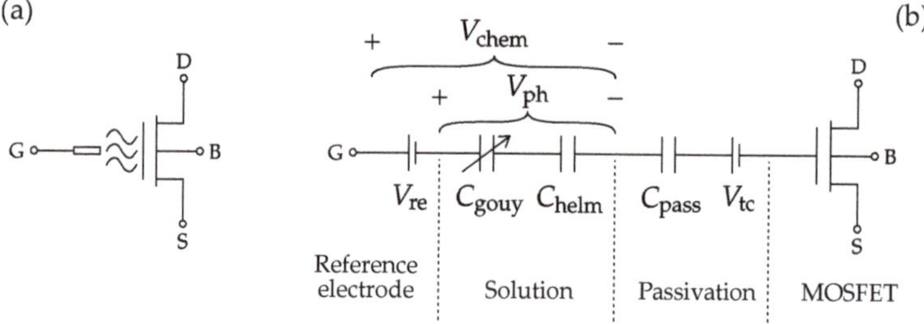

Figure 19. Ion-selective field-effect transistor (ISFET) symbol (**a**) and equivalent electrical model (**b**).

The corresponding pH-dependent shift on the equivalent ISFET threshold voltage can be expressed as:

$$\Delta V_{TH} = \gamma + \alpha S_N \, pH, \tag{26}$$

where γ is a chemical constant, S_N stands for the theoretical Nernstian pH sensitivity (typ. 59 mV/pH) and α counts for the sensitivity deviations due to the double-layer capacitance [169].

Several specific AFE circuits have been proposed for the ISFET from the very beginning of its birth. However, the last decade has experienced an exponential growth in interface topologies thanks to the emergence of integrated arrays of ISFETs for lab-on-a-chip instruments [170] with pixel-like CMOS read-out circuits monolithically attached one-by-one to their corresponding ISFET sensors. In this review, two types of ISFET interfaces are presented according to potentiometric and amperometric read-out methods.

Figure 20 shows a summary of the common ISFET potentiometric configurations, where V_{ref}, V_A and d_o stand for the RE potential, the AFE output and the digital output of the overall smart ISFET front-end, respectively. The most classic solution is probably the constant-I_D constant-V_{DS} dual-OpAmp topology [171] of Figure 20a due to its excellent linearity in all regions of MOSFET operation, from weak to strong inversion and from conduction to saturation. Its principle of operation can be explained as follows: the current sink I_2 ensures constant-I_D for the ISFET device M_1, which is configured here as a gate-to-source voltage follower; the unity-gain OpAmp A_2 provides the low-impedance copy V_A of the ISFET source voltage for further A/D conversion; I_1 and R_1 define the selected constant V_{DS} bias, which is then applied to M_1 by the unity-gain OpAmp A_1. Based on the same strategy, a single-OpAmp topology is presented in Figure 20b. In both cases, a separated substrate (i.e., local well) for the ISFET is needed in order to avoid any CMOS technology dependency. For example, if the floating-gate MOSFET M_1 is biased in strong inversion forward saturation, the EKV model [172] returns the following large-signal DC characteristic:

$$I_D = \frac{\beta}{2n} \left(V_{GB} - V_{TH} - n V_{SB} \right)^2 \left(1 + \lambda V_{DS} \right), \tag{27}$$

where β is the current factor (including device aspect ratio), n is the subthreshold slope and λ is the channel-length modulation parameter. Clearly, any variation of the gate voltage ΔV_{GB} (or alternatively of the threshold voltage ΔV_{TH}) in Equation (27) due to pH will cause, at constant I_D and V_{DS}, the corresponding change in the source voltage $\Delta V_{SB} = \Delta V_{TH}/n$. This subthreshold slope dependency can be avoided by forcing $V_{SB} = 0$ (i.e., local well). When this technology option is not available, like in twin-well CMOS processes, the alternative topology [173] of Figure 20c is more suitable. In this case, a matched MOSFET M_2 is introduced to form a differential pair with ISFET M_1. If both devices are biased at the same constant current (i.e., $I_1 = I_2 = I_3/2$), the high-gain negative feedback will force both devices to share the same V_{DS} and consequently, from Equation (27), M_2 will follow any voltage variation at the floating gate of M_1 even under substrate effects ($V_{SB} \neq 0$).

The common approach of the ISFET interface circuits presented in Figure 20 is the usage of the floating-gate MOS transistor as a voltage follower. However, the true nature of the MOSFET is of a transconductor, which is a voltage-controlled current source, which opens up the possibility for ISFET amperometric topologies. Compared to potentiometric techniques, current-mode read out can enjoy those circuit design techniques described in Section 4.2. In addition, signals in the current domain can be directly filtered or employed in modulation schemes with the inclusion of CMOS integrators. On the other hand, ISFET amperometric circuit strategies tend to suffer from the nonlinear I/V MOS device characteristics.

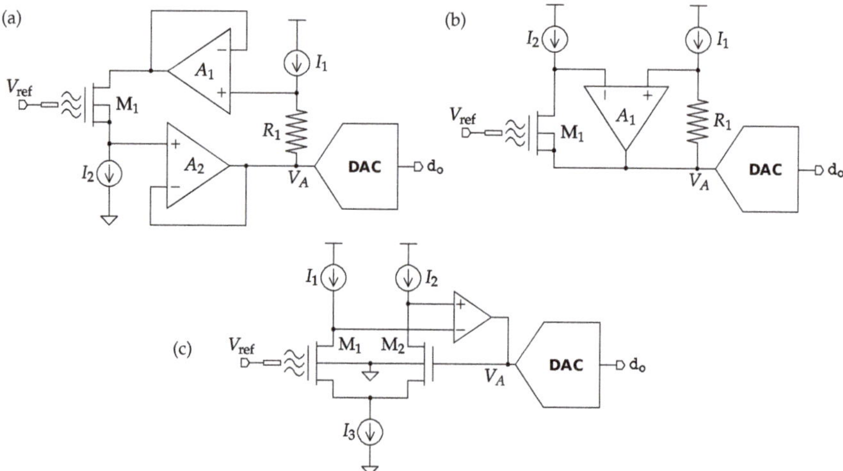

Figure 20. Potentiometric ISFET interfaces: (**a**) constant-I_D constant-V_{DS} classic dual-OpAmp and (**b**) single-OpAmp topologies, and (**c**) constant-I_D OpAmp follower.

Figure 21 presents two common ISFET amperometric interfaces. In both examples, the floating-gate MOSFET is biased in its strong inversion triode region by selecting $V_{bias} \leq V_{Dsat}$. According to the EKV model [172],

$$I_D = \beta \left[V_{GB} - V_{TH} - \frac{n}{2} \left(V_{DB} + V_{SB} \right) \right] V_{DS}. \tag{28}$$

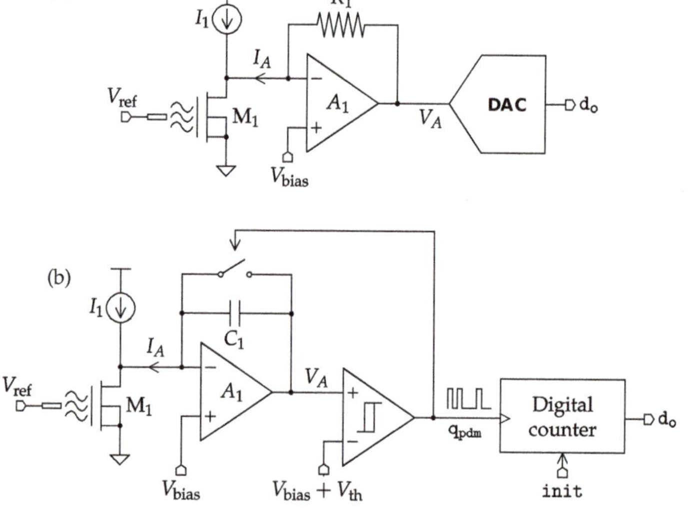

Figure 21. Amperometric ISFET readout interfaces: (**a**) classic resistive TIA analog front-end and (**b**) asynchronous pulse-density modulation (PDM) analog-to-digital converter with capacitive-TIA.

Thus, the current-mode signal I_A shows a linear behaviour respect to any change in the floating-gate voltage of the ISFET M_1 with a transconductance gain βV_{bias}. In the classic topology of Figure 21a, I_A is converted back to the voltage domain V_A through the resistive TIA $R_1 A_1$ for further A/D conversion [174]. A completely different approach is followed in the interface of Figure 21b, where the same current-domain signal is directly employed for the A/D conversion. Its principle of operation can be summarized as follows: I_A is integrated through the CTIA C_1A_1; the resulting signal V_A is then 1-bit quantized at q_{pdm} by the comparator with hysteresis according to the voltage threshold V_{th}; the same digital signal is fed back to the CTIA as the trigger to reset the continuous-time analog integration. As a result, q_{pdm} shows a pulse density modulation (PDM) with respect to I_A (pH). The only remaining signal processing step to complete the A/D conversion of I_A is some sort of digital low-pass filtering, which is implemented in Figure 21b by a simple digital counter playing the role of a first-order low-pass filter. Apart from this asynchronous integrated-and-fire strategy, other I_A modulation schemes are also possible for the practical implementation of the A/D conversion, like pulse width modulation (PWM) [175].

4.4. Impedance Spectroscopy Interfaces

An EIS circuit must be capable of measuring how the complex small-signal impedance of an electrolyte/electrode interface varies with frequency. The aim is determining the main features of the typical curve shown in Figure 5 (right). Simpler systems designed to measure only the capacitive component at fixed frequencies are sometimes used for their simplicity when the goal is finding the coverage of the functionalized working electrode by the target molecules [176] or detecting cellular activity [177]. Classification of the EIS systems proposed in the literature can be made on the basis of the diagram of Figure 22.

The stimulus generator produces a reference waveform, which is applied to the EIS interface through a proper electrode configuration. The response is then processed to extract the complex impedance (real and imaginary part). Two main working principles are possible: the fast Fourier transform (FFT) and the frequency response analysis (FRA). The former consists of stimulating the EIS interface with a waveform having a broad spectrum, such as a single pulse or a square waveform. The response is converted to digital and FFT is performed in order to extract the required dependence of the impedance on frequency. The FFT approach is not the optimal solution for compact, single-chip EIS systems, due to the requirement of complex digital-signal processing (DSP) functionalities.

With the FRA approach, the stimulus is sinusoidal and the response is analyzed in order to extract the complex impedance at the stimulus frequency. The impedance curve is then obtained through a frequency scan over the interval of interest. In both cases, the AC stimulus has a small magnitude (tens of mV) and it must be superimposed on a larger DC component, which is used to activate the desired redox processes. When the stimulus is applied as a voltage, imposed between the reference electrode and the working electrode, the current flowing through the working electrode is acquired. Current stimulation is not common in EIS systems but is frequently used in bio-impedance measurement devices [178–180], where it is not important to set a precise DC bias.

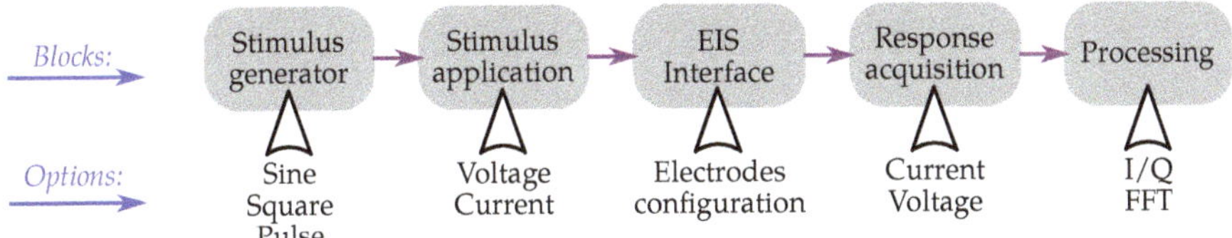

Figure 22. Fundamental elements of an electrochemical impedance spectroscopy (EIS) interface.

The most popular approach for the development of compact EIS systems is using voltage stimulation combined with FRA operating principles. Accurate EIS approaches involve using the typical three-electrode potentiostat configuration (three-electrode connection), where the voltage stimulus is imposed across the RE and WEs through the action of a CE, as in voltammetry systems. However, due to its simplicity, the two-electrode is preferred for large sensor arrays where a large common CE is coupled with an array of smaller WEs (usually microelectrodes) [181,182]. In such experiments, the current levels at WEs are low enough that, even if summed up at the large common CE, its potential drift is negligible and a sufficiently precise potentiostatic control is established.

FRA requires a sinusoidal stimulus; square wave AC stimulation is generally considered not adequate due to the excessively high content of harmonics that would alter the measurement [181,183]. Due to the need for sweeping the frequency across intervals several decades wide, classical RC or LC oscillators are not adequate for generating the stimulus. The preferred solution is synthesizing the sinusoid by means of a conventional *R-2R* [181] or resistor-string [179] DAC, starting from a set of samples.

High spectral purity requires high resolution DACs and a large number of samples per cycle, which are factors that lead to an increase of power consumption and architectural complexity. An alternative approach is starting from a square waveform and filtering it to select only the fundamental component. Due to the wide frequency interval, time-continuous filtering is not feasible. In [184] synchronization of the filter-bandwidth with the stimulus frequency is obtained by using a switched capacitors 5th order Chebyshev type-I filter clocked at a multiple of the input frequency.

The FRA approach requires that the response of the interface under test (IUT) be processed to extract the impedance real and imaginary parts. The IUT response is either a voltage or a current. The straightforward solution is shown in Figure 23, where the IUT response is multiplied by a stimulus time reference (STR), i.e., a signal synchronous with the input stimulus. Multiplication for the STR and for the STR delayed by one-fourth of period ($\pi/2$ phase delay) produces two signals whose DC components are proportional to the in-phase (I) and quadrature (Q) components of the IUT response, respectively. For a voltage stimulus and current response, the I and Q signals are proportional to the real and imaginary parts of the electrochemical cell admittance.

The scheme of Figure 23 is frequently implemented by all-analog circuits, where the multipliers are formed by switch matrices and the STR is a square waveform. Even with this type of simplification, full integration of the scheme in Figure 23 can be quite challenging, due to the difficulty of designing low pass filters (LPF) with sub-Hz cut-off frequencies using the pF-range capacitors available on-chip. In Ref. [179], a Gm-C filter that relies on external capacitors is employed, whereas an external filter/amplifier is used in [182].

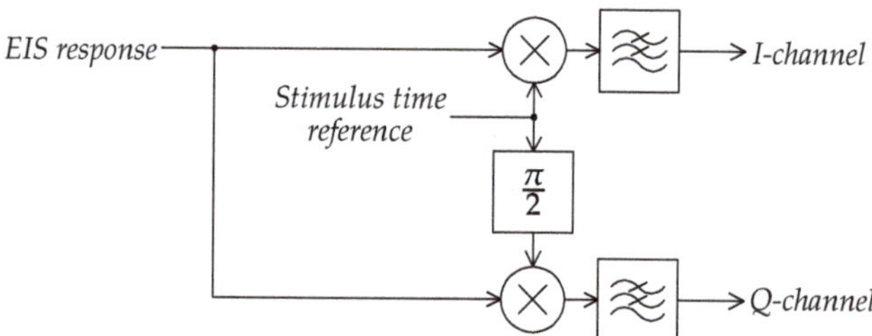

Figure 23. I/Q demodulation of the EIS response.

An alternative solution, corresponding to multiplying the EIS interface response by a discretized sinusoidal reference, is proposed in [181]. Such a result is obtained by digitizing the response by means of a dual-slope ADC, whose integration time is digitally modulated by the sinusoidal STR. An advantage of this solution is that a digital signal is produced and no further filtering is required. Another mixed analog-digital approach based on a pass-band delta-sigma ADC is sketched in [185], while all-digital demodulation solutions can be obtained by processing the digitized EIS interface response in the digital domain to calculate the single-frequency the Discrete Fourier Transform (DFT). Such a solution [186] represents an FFT/FRA hybrid approach and allows for relaxed DSP requirements with respect to pure FFT based approaches, but sinusoidal excitation and the execution of a frequency scan are necessary as in FRA. As an alternative to extracting the I and Q components, separate magnitude and phase estimation has been demonstrated to be feasible [187].

4.5. Readout Interfaces for Bio-Fuel Cells

Amongst the latest trends, sensing based on biological fuel cells (BFCs) is a promising approach for wearables. Different definitions of a BFC may apply [188]; however, in the present context, a BFC can be defined as a device that utilizes at least one biogenic catalyst, such as enzymes, microbes or microorganisms, in order to assimilate a fuel and an oxidant to generate electric power. BFCs are able to generate charge and simultaneously store it in large double layer and/or pseudo-capacitance structures around the electrodes. In a BFC, as in any generic fuel cell, the amount of available power P_{out} is set by the concentration of the fuel. At the same time, a BFC based on glucose oxidase (GOx) or lactate oxidase (LOx) may use respectively glucose or lactate, naturally present in sweat, as fuels [189]. In turn, since glucose and lactate are markers of metabolic activity, they can be leveraged to provide both information and energy for information readout, digitalization and transmission, giving yield to a fully self-powered smart wearable. Power densities in the order of 1 mW/cm^2 have been recently demonstrated for on-body lactate BFCs [78,190].

From the electrical point of view, a BFC behaves as a voltage source V_{cell} able to sustain a certain amount of current I_{cell}, whose intensity depends on the concentration of the fuel. Schematic polarization curves are shown in Figure 24. For an unloaded BFC, i.e., in the open-circuit condition, V_{cell} is at a maximum and it is usually on the order of 0.3–0.5 V. Once I_{cell} starts to be provided to a load, V_{cell} drops in a resistive-like manner up to the overloading condition of the BFC, where V_{cell} drops dramatically to zero. The power versus cell voltage characteristic reveals a maximum power point (MPP) which optimizes the power harvested from the BFC. However from the CMOS point of view, operation at MPP, which falls around $V_{cell} \approx 0.2$ V, is extremely challenging and different strategies may be adopted.

A possible solution is depicted in Figure 25a [191]. This system uses energy supplied by the BFC and provides glucose/lactate concentration readout and RF data-transmission in the UHF band. The system operates in two phases: during φ_{charge}, the BFC is charging an external ceramic capacitor C_{ext}, employed as auxiliary energy storage element. Provided that φ_{charge} is long enough, the transient currents finally approach zero, meaning that the available supply voltage for circuit operation will be the open circuit value of V_{cell}. During $\varphi_{readout}$, the circuit is turned on and supplied by the harvested energy in C_{ext}. In this phase, an approximated matched load R_{MPP} is connected to the BFC in order to track the MPP. This need is justified by the authors in [191] by experimental evidence disclosing a linear relationship between the P_{out} and analyte concentration when operating at MPP.

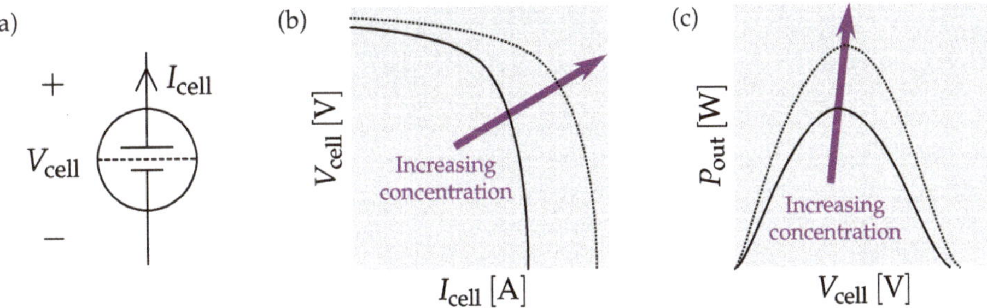

Figure 24. Bio-fuel cell symbol (**a**) and schematic polarization curves: V_{cell} versus I_{cell} (**b**) and output power P_{out} versus V_{cell} (**c**).

A more compact solution for BFC interfacing is proposed in [192] and it is reported in Figure 25b. Here, an all-digital approach is followed comprising a supply-controlled ring oscillator, a pulse shaper and an impulse-based transmitter for inductive-coupling. Linearity performances are worse with respect to the previous solution; however, only three connections to external components are required, with clear advantages to the heterogeneous integration of the chip onto a flexible substrate. Using the zero-Vth CMOS transistors, the test chip is able to function with V_{cell} as low as 0.23 V.

While both reviewed works present pioneer solutions, it is worth noting that [191] fails to address the physiological normal ranges of Table 1. On the other hand, Ref. [192] is interfaced to a BFC fuelled by fructose, which is an analyte not present in human body fluids. Thus, BFC interfacing is still a green field and many more solutions are expected to be proposed over the next years.

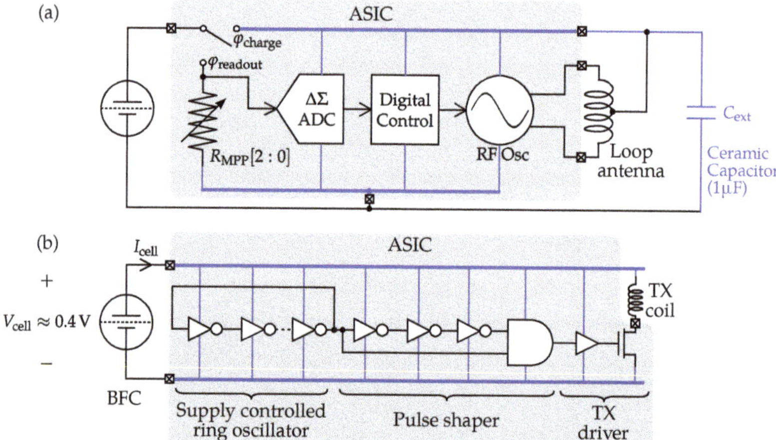

Figure 25. CMOS interface for self-powered fuel cell sensors. (a) two-phase operation approach; (b) free-running oscillator approach.

5. Conclusions

We have reviewed the state-of-the-art CMOS interfaces for non-invasive monitoring of physiological parameters through electro-analytical techniques in body fluids. Such integrated circuits constitute the first stage of the signal acquisition and processing chain in a system comprising (bio)electrochemical sensors to be embedded in mechanically flexible substrates. While constrained to be low-power and to have a small footprint, these front-end circuits bridge the gap between the transduced low-level signals and Cloud-connected components such as smartphones, smartwatches or dedicated WPAN links. We coined the term IoW to identify these kinds of smart systems that find application in health, sports, safety at work, defence and law enforcement. With the aim to provide a self-contained framework, an overview of electro-analytical methods including potentiometric, amperometric and impedimetric sensing has been given. Challenges in CMOS circuit design for potentiometric sensing, both by the means of ISE and ISFET sensing elements, have been stated and analysed. CMOS potentiostats for amperometric sensing, popular thanks to its customization to a vast amount of target analytes, have been extensively discussed, comprising the most recent compact solutions. More complex systems such as those based on EIS need sophisticated analog, mixed-signal and digital signal processing. This implies a penalty in terms of silicon area and power consumption directly impacting wearable cost and portability. Finally, BFC-based smart systems, still in their embryonic stage, have been identified as a promising emerging field thanks to their self-powering characteristics.

Author Contributions: Conceptualization, M.D. and F.J.d.C.; Methodology, M.D.; Supervision, F.J.d.C., P.B., F.S.-G.; Writing and Editing, all authors.

Funding: The research leading to these results has received funding from the People Programme (Marie Curie Actions) of the Seventh Framework Programme of the European Union (FP7/2007–2013) under REA Grant No. 600388 (TECNIOspring programme: TECSPR16-1-0056), and from the Agency for Business Competitiveness of the Government of Catalonia, ACCIÓ. It has also received funding from "PRODUCTE 2016" grants (PROD-00114) from the knowledge industry program run by the Universities and Research secretariat from the Catalan regional government.

Conflicts of Interest: The authors declare no conflict of interest. The founding sponsors had no role in the design of the study; in the collection, analyses, or interpretation of data; in the writing of the manuscript, or in the decision to publish the results.

References

1. Liu, X.; Li, L.; Mason, A.J. *Handbook of Bioelectronics*; Carrara, S., Iniewski, K., Eds.; Cambridge University Press: Cambridge, UK, 2015; ISBN 978-1-107-04083-0.

2. Ghafar-Zadeh, E. Wireless Integrated Biosensors for Point-of-Care Diagnostic Applications. *Sensors* **2015**, *15*, 3236–3261, doi:10.3390/s150203236.

3. Li, H.; Liu, X.; Li, L.; Mu, X.; Genov, R.; Mason, A.J. CMOS Electrochemical Instrumentation for Biosensor Microsystems: A Review. *Sensors* **2017**, *17*, 74, doi:10.3390/s17010074.

4. Market Research Engine. Available online: www.marketresearchengine.com/wearable-devices-market (accessed on 21 January 2019).

5. Lymberis, A. Feedback from Stakeholders on the Smart Wearables Reflection and Orientation Paper. *News Article*. Available online: https://ec.europa.eu/digital-single-market/en/news/feedback-stakeholders-smart-wearables-reflection-and-orientation-paperec.europa.eu/digital-single-market/en/news/feedback-stakeholders-smart-wearables-reflection-and-orientation-paper (accessed on 21 January 2019).

6. Bariya, M.; Nyein, H.Y.Y.; Javey, A. Wearable sweat sensors. *Nat. Electron.* **2018**, *1*, 160–171, doi:10.1038/s41928-018-0043-y.

7. Kim, J.; Jeerapan, I.; Sempionatto J.R.; Barfidokht, A.; Mishra, R.K.; Campbell, A.S.; Hubble, L.J.; Wang, J. Wearable Bioelectronics: Enzyme-Based Body-Worn Electronic Devices. *Acc. Chem. Res.* **2018**, *51*, 2820–2828, doi:10.1021/acs.accounts.8b00451.

8. Ngamchuea, K.; Chaisiwamongkhol, K.; Batchelor-McAuleya C.; Compton, R.G. Chemical analysis in saliva and the search for salivary biomarkers—A tutorial review. *Analyst* **2018**, *143*, 81–99, doi:10.1039/c7an01571b.

9. Ngamchuea, K.; Chaisiwamongkhol, K.; Batchelor-McAuleya C.; Compton, R.G. Correction: Chemical analysis in saliva and the search for salivary biomarkers—A tutorial review. *Analyst* **2018**, *143*, 777–783, doi:10.1039/c7an90101a.

10. Farandos, N.M.; Yetisen, A.K.; Monteiro, M.J.; Lowe, C.R.; Yun, S.H. Contact Lens Sensors in Ocular Diagnostics. *Adv. Healthc. Mater.* **2014**, *4*, 792–810, doi:10.1002/adhm.201400504.

11. Selvam, A.P.; Muthukumar, S.; Kamakoti, V.; Prasad, S. A wearable biochemical sensor for monitoring alcohol consumption lifestyle through Ethyl glucuronide (EtG) detection in human sweat. *Sci. Rep.* **2016**, *6*, 23111, doi:10.1038/srep23111.

12. Kim, J.; Sempionatto, J.R.; Imani, S.; Hartel, M.C.; Barfidokht, A.; Tang, G.; Campbell, A.S.; Mercier, P.P.; Wang, J. Simultaneous Monitoring of Sweat and Interstitial Fluid Using a Single Wearable Biosensor Platform. *Adv. Sci.* **2018**, *5*, 1800880, doi:10.1002/advs.201800880.

13. Sempionatto, J.R.; Martin, A.; García-Carmona, L.; Barfidokht, A.; Kurniawan, J.F.; Moreto, J.R.; Tang, G.; Shin, A.; Liu, X.; Escarpa, A.; et al. Skin-Worn Soft Microfluidic Potentiometric Detection System. *Electroanalysis* **2018**, doi:10.1002/elan.201800414.

14. Kim, S.B.; Lee, K.H.; Raj, M.S.; Lee, B.; Reeder, J.T.; Koo, J.; Hourlier-Fargette, A.; Bandodkar, A.J.; Won, S.M.; Sekine, Y.; et al. Soft, Skin-Interfaced Microfluidic Systems with Wireless, Battery-Free Electronics for Digital, Real-Time Tracking of Sweat Loss and Electrolyte Composition. *Small* **2018**, *14*, e1802876, doi:10.1002/smll.201802876.

15. Anastasova, S.; Crewther, B.; Bembnowicz, P.; Curto, V.; Ip, H.M.; Rosa, B.; Yang, G.Z. A wearable multisensing patch for continuous sweat monitoring. *Biosens. Bioelectron.* **2017**, *93*, 139–145, doi:10.1016/j.bios.2016.09.038.

16. Burtis, C.A.; Bruns, D.E. *Tietz Fundamentals of Clinical Chemistry and Molecular Diagnostics*, 7th Ed.; Elsevier Health Sciences: Amsterdam, The Netherlands, 2015; ISBN 978-1-4557-4165-6.

17. Baker, L.B. Sweating Rate and Sweat Sodium Concentration in Athletes: A Review of Methodology and Intra/Interindividual Variability. *Sports Med.* **2017**, *47*, 111–128, doi:10.1007/s40279-017-0691-5.

18. Mitsubayashi, K.; Suzuki, M.; Tamiya, E.; Karube, I. Analysis of metabolites in sweat as a measure of physical condition. *Anal. Chim. Acta* **1994**, *289*, 27–34, doi:10.1016/0003-2670(94)80004-9.

19. Sakharov, D.A.; Shkurnikov, M.U.; Vagin, M.Y.; Yashina, E.I.; Karyakin, A.A.; Tonevitsky, A.G. Relationship between lactate concentrations in active muscle sweat and whole blood. *Bull. Exp. Biol. Med.* **2010**, *150*, 83–85.

20. Pribil, M.M.; Laptev, G.U.; Karyakina, E.E.; Karyakin, A.A. Noninvasive Hypoxia Monitor Based on Gene-Free Engineering of Lactate Oxidase for Analysis of Undiluted Sweat. *Anal. Chem.* **2014**, *86*, 5215–5219, doi:10.1021/ac501547u.

21. Van Haeringen, N.J. Clinical biochemistry of tears. *Surv. Ophthalmol.* **1981**, *26*, 84–96.

22. Carreño, E.; Enríquez-de-Salamanca, A.; Tesón, M.; García-Vázquez, C.; Stern, M.E.; Whitcup, S.M.; Calonge, M. Cytokine and chemokine levels in tears from healthy subjects. *Acta Ophthalmol.* **2010**, *88*, e250–e258, doi:10.1111/j.1755-3768.2010.01978.x.

23. Bruen, D.; Delaney, C.; Florea, L.; Diamond, D. Glucose Sensing for Diabetes Monitoring: Recent Developments. *Sensors* **2017**, *17*, 1866, doi:10.3390/s17081866.

24. Abelson, M.B.; Udell, I.J.; Weston, J.H. Normal human tear pH by direct measurement. *Arch. Ophthalmol.* **1981**, *99*, 301.

25. Tékus, É.; Kaj, M.; Szabó, E.; Szénási, N.; Kerepesi, I.; Figler, M.; Gábriel, R.; Wilhelm, M. Comparison of blood and saliva lactate level after maximum intensity exercise. *Acta Biol. Hung.* **2012**, *63*, 89–98, doi:10.1556/ABiol.63.2012.Suppl.1.9.

26. Hagen, T.; Korson, M.; Wolfsdorf J. Urinary lactate excretion to monitor the efficacy of treatment of type I glycogen storage disease. *Mol. Genet. Metab.* **2000**, *70*, 189–195, doi:10.1006/mgme.2000.3013.

27. Steckl, A.J.; Ray, P. Stress Biomarkers in Biological Fluids and Their Point-of-Use Detection. *ACS Sens.* **2018**, *3*, 2025–2044, doi:10.1021/acssensors.8b00726.

28. Ginsberg, B.H.; An overview of minimally invasive technologies. *Clin. Chem.* **1992**, *38*, 1596–1600.

29. World Health Organization, Diabetes Programme. Available online: https://www.who.int/diabetes/en/ (accessed on 21 January 2019).

30. Wilson, R.; Turner, A.P.F. Glucose oxidase: An ideal enzyme. *Biosens. Bioelectron.* **1992**, *7*, 165–185.

31. Wang, J. Electrochemical glucose biosensors. *Chem. Rev.* **2008**, *108*, 814–825, doi:10.1021/cr068123a.

32. Toghill, K.E.; Compton, R.G. Electrochemical non-enzymatic glucose sensors: A perspective and an evaluation. *Int. J. Electrochem. Sci.* **2010**, *5*, 1246–1301.

33. Kim, J.; Campbell, A.S.; Wang, J. Wearable non-invasive epidermal glucose sensors: A review. *Talanta* **2018**, *177*, 163–170, doi:10.1016/j.talanta.2017.08.077.

34. Gifford, R. Continuous Glucose Monitoring: 40 Years, What We've Learned and What's Next. *ChemPhysChem* **2013**, *14*, 2032–2044, doi:10.1002/cphc.201300172.

35. Clark, L.C.; Lyons, C. Electrode systems for continuous monitoring in cardiovascular surgery. *Ann. N. Y. Acad. Sci.* **1962**, *102*, 29–45.

36. Tierney, M.J.; Tamada, J.A.; Potts, R.O.; Jovanovic, L.; Garg, S. Clinical evaluation of the GlucoWatch biographer: A continual, non-invasive glucose monitor for patients with diabetes. *Biosens. Bioelectron.* **2001**, *16*, 621–629.

37. FreeStyle Libre. Available online: https://www.freestylelibre.com (accessed on 21 January 2019).

38. Hoss, U.; Budiman, E.S. Factory-Calibrated Continuous Glucose Sensors: The Science Behind the Technology. *Diabetes Technol. Ther.* **2017**, *19*, S-44–S-50, doi:10.1089/dia.2017.0025.

39. James, D.A.; Petrone, N. *Sensors and Wearable Technologies in Sport: Technologies, Trends and Approaches for Implementation*; Future Direction Sereires; Springer: Singapore, 2016; p. 49, ISBN 978-981-10-0992-1.

40. Rossi, A.; Pappalardo, L.; Cintia, P.; Iaia, F.M.; Fernàndez, J.; Medina, D. Effective injury forecasting in soccer with GPS training data and machine learning. *PLoS ONE* **2018**, *13*, e0201264, doi:10.1371/journal.pone.0201264.

41. Seshadri, D.R.; Drummond, C.; Craker, J.; Rowbottom, J.R.; Voos, J.E. Wearable Devices for Sports: New Integrated Technologies Allow Coaches, Physicians, and Trainers to Better Understand the Physical Demands of Athletes in Real Time. *IEEE Pulse* **2017**, *8*, 38–43, doi:10.1109/MPUL.2016.2627240.

42. Peng, R.; Sonner, Z.; Hauke, A.; Wilder, E.; Kasting, J.; Gaillard, T.; Swaille, D.; Sherman, F.; Mao, X.; Hagen, J.; et al. A new oil/membrane approach for integrated sweat sampling and sensing: sample volumes reduced from μL's to nL's and reduction of analyte contamination from skin. *Lab Chip* **2016**, *16*, 4415–4423, doi:10.1039/c6lc01013j.

43. Nyein, H.Y.Y.; Tai, L.C.; Ngo, Q.P.; Chao, M.; Zhang, G.B.; Gao, W.; Bariya, M.; Bullock, J.; Kim, H.; Fahad, H.M.; et al. A Wearable Microfluidic Sensing Patch for Dynamic Sweat Secretion Analysis. *ACS Sens.* **2018**, *3*, 944–952, doi:10.1021/acssensors.7b00961.

44. Choi, J.; Xue, Y.; Xia, W.; Ray, T.R.; Reeder, J.T.; Bandodkar, A.J.; Kang, D.; Xu, S.; Huang, Y.; Rogers, J.A. Soft, skin-mounted microfluidic systems for measuring secretory fluidic pressures generated at the surface of the skin by eccrine sweat glands. *Lab Chip* **2017**, *17*, 2572–2580, doi:10.1039/c7lc00525c.

45. Koh, A.; Kang, D.; Xue, Y.; Lee, S.; Pielak, R.M.; Kim, J.; Hwang, T.; Min, S.; Banks, A.; Bastien, P.; et al. A soft, wearable microfluidic device for the capture, storage, and colorimetric sensing of sweat. *Sci. Transl. Med.* **2016**, *8*, 366ra165, doi:10.1126/scitranslmed.aaf2593.

46. Garcia, S.O.; Ulyanovaa, Y.V.; Figueroa-Teranb, R.; Bhatta, K.H.; Singhala, S.; Atanassov, P. Wearable Sensor System Powered by a Biofuel Cell for Detection of Lactate Levels in Sweat. *ECS J. Solid State Sci. Technol.* **2016**, *5*, M3075–M3081, doi:10.1149/2.0131608jss.

47. Toft, A.D.; Jensen, L.B.; Bruunsgaard, H.; Ibfelt, T.; Halkjær-Kristensen, J.; Febbraio, M.; Pedersen, B.K. Cytokine response to eccentric exercise in young and elderly humans. *Am. J. Physiol. Cell Physiol.* **2002**, *283*, C289–C295, doi:10.1152/ajpcell.00583.2001.

48. Marques-Deak, A.; Marques-Deak, A.; Cizza, G.; Eskandari, F.; Torvik, S.; Christie, I.C.; Sternberg, E.M.; Phillips, T.M. Measurement of cytokines in sweat patches and plasma in healthy women: Validation in a controlled study. *J. Immunol. Methods* **2006**, *315*, 99–109, doi:10.1016/j.jim.2006.07.011.

49. Willner, I.; Zayats, M. Electronic aptamer-based sensors. *Angew. Chem.* **2007**, *46*, 6408–6418, doi:10.1002/anie.200604524.

50. Kumar, L.S.S.; Wang, X.; Hagen, J.; Naik, R.; Papautskya, I.; Heikenfeld, J. Label free nano-aptasensor for interleukin-6 in protein-dilute bio fluids such as sweat. *Anal. Methods* **2016**, *8*, 3440–3444, doi:10.1039/C6AY00331A.

51. Morgan, R.M.; Patterson, M.J.; Nimmo, M.A. Acute effects of dehydration on sweat composition in men during prolonged exercise in the heat. *Acta Physiol. Scand.* **2004**, *182*, 37–43, doi:10.1111/j.1365-201X.2004.01305.x.

52. Cheuvront, S.N.; Kenefick, R.W. Dehydration: Physiology, assessment, and performance effects. *Compr. Physiol.* **2014**, *4*, 257–285, doi:10.1002/cphy.c130017.

53. Moran, P.; Prichard, J.G.; Ansley, L.; Howatson, G. The influence of blood lactate sample site on exercise prescription. *J. Strength Cond. Res.* **2012**, *26*, 563–567, doi:10.1519/JSC.0b013e318225f395.

54. Sonner, Z.; Wilder, E.; Heikenfeld, J.; Kasting, G.; Beyette, F.; Swaile, D.; Sherman, F.; Joyce, J.; Hagen, J.; Kelley-Loughnane, N.; et al. The microfluidics of the eccrine sweat gland, including biomarker partitioning, transport, and biosensing implications. *Biomicrofluidics* **2015**, *9*, 031301, doi:10.1063/1.4921039.

55. Wearable Devices Used for Industrial Applications, Available online: https://vandrico.com/wearables/device-categories/application/industrial (accessed on 21 January 2019).

56. Cone, E.J.; Hillsgrove, M.J.; Jenkins, A.J.; Keenan, R.M.; Darwin, W.D. Sweat testing for heroin, cocaine, and metabolites. *J. Anal. Toxicol.* **1994**, *18*, 298–305.

57. Mishra, R.K.; Hubble, L.J.; Martín, A.; Kumar, R.; Barfidokht, A.; Kim, K.; Musameh, M.M.; Kyratzis, I.L.; Wang, J. Wearable Flexible and Stretchable Glove Biosensor for On-Site Detection of Organophosphorus Chemical Threats. *ACS Sens.* **2017**, *2*, 553–561, doi:10.1021/acssensors.7b00051.

58. Mishra, R.K.; Barfidokht, A.; Karajic, A.; Sempionatto, J.R.; Wang, J.; Wang, J. Wearable potentiometric tattoo biosensor for on-body detection of G-type nerve agents simulants. *Sens. Actuator B Chem.* **2018**, *273*, 966–972, doi:10.1016/j.snb.2018.07.001.

59. Mishra, R.K.; Martín, A.; Nakagawa, T.; Barfidokht, A.; Lu, X.; Sempionatto, J.R.; Lyu, K.M.; Karajic, A.; Musameh, M.M.; Kyratzis, I.L.; et al. Detection of vapor-phase organophosphate threats using wearable conformable integrated epidermal and textile wireless biosensor systems. *Biosens. Bioelectron.* **2018**, *101*, 227–234, doi:10.1016/j.bios.2017.10.044.

60. Huang, Y.; Tzeng, T.; Lin, T.; Huang, C.; Yen, P.; Kuo, P.; Lin, C.; Lu, S. A Self-Powered CMOS Reconfigurable Multi-Sensor SoC for Biomedical Applications. *IEEE J. Solid-State Circuits* **2014**, *49*, 851–866, doi:10.1109/JSSC.2013.2297392.

61. Sun, A.; Venkatesh, A.G.; Hall, D.A. A Multi-Technique Reconfigurable Electrochemical Biosensor: Enabling Personal Health Monitoring in Mobile Devices. *IEEE Trans. Biomed. Circuits Syst.* **2016**, *10*, 945–954, doi:10.1109/TBCAS.2016.2586504.

62. Rose, D.P.; Ratterman, M.E.; Griffin, D.K.; Hou, L.; Kelley-Loughnane, N.; Naik, R.R.; Hagen, J.A.; Papautsky, I.; Heikenfeld, J.C. Adhesive RFID Sensor Patch for Monitoring of Sweat Electrolytes. *IEEE Trans. Biomed. Circuits Syst.* **2015**, *62*, 1457–1465, doi:10.1109/TBME.2014.2369991.

63. Beni, V.; Nilsson, D.; Arven, P.; Norberg, P.; Gustafsson, G.; Turner, A.P.F. Printed Electrochemical Instruments for Biosensors. *ECS J. Solid State Sci. Technol.* **2015**, *4*, S3001–S3005, doi:10.1149/2.0011510jss.

64. Gao, W.; Emaminejad, S.; Nyein, Y.Y.H.; Challa, S.; Chen, K.; Peck, A.; Hossain, M.F.; Ota, H.; Shiraki, H.; Kiriya, D.; et al. Fully integrated wearable sensor arrays for multiplexed in situ perspiration analysis. *Nat. Lett.* **2016**, *529*, 509–514, doi:10.1038/nature16521.

65. Aller-Pellitero, M.; Guimerà, A.; Kitsara, M.; Villa, R.; Rubio, C.; Lakard, B.; Doche, M.L.; Hihnc, J.Y.; del Campo, F.J. Quantitative self-powered electrochromic biosensors. *Chem. Sci.* **2017**, *8*, 1995–2002, doi:10.1039/C6SC04469G.

66. Jung, Y.; Park, H.; Park, J.; Noh, J.; Choi, Y.; Jung, M.; Jung, K.; Pyo, M.; Chen, K.; Javey, A.; et al. Fully printed flexible and disposable wireless cyclic voltammetry tag. *Sci. Rep.* **2015**, *5*, 8105. doi:10.1038/srep08105.

67. Chang, J.S.; Facchetti, A.F.; Reuss, R. A Circuits and Systems Perspective of Organic/Printed Electronics: Review, Challenges, and Contemporary and Emerging Design Approaches. *IEEE Trans. Emerg. Sel. Top. Circuits Syst.* **2017**, *7*, 7–26, doi:10.1109/JETCAS.2017.2673863.

68. Shiwaku, R.; Matsui, H.; Nagamine, K.; Uematsu, M.; Mano, T.; Maruyama, Y.; Nomura, A.; Tsuchiya, K.; Hayasaka, K.; Takeda, Y.; Fukuda, T.; Kumaki, D.; Tokito S. A printed Organic Circuit System for Wearable Amperometric Electrochemical Sensors. *Sci. Rep.* **2018**, *8*, 6368, doi:10.1038/s41598-018-24744-x.

69. Liao, Y.; Yao, H.; Lingley, A.; Parviz, B.; Otis, B.P. A 3-μW CMOS Glucose Sensor for Wireless Contact-Lens Tear Glucose Monitoring. *IEEE J. Solid-State Circuits* **2012**, *47*, 335–344, doi:10.1109/JSSC.2011.2170633.

70. Huang, W.; Deb, S.; Seo, Y.; Rao, S.; Chiao, M.; Chiao, J.C. A Passive Radio-Frequency pH-Sensing Tag for Wireless Food-Quality Monitoring. *IEEE Sens. J.* **2012**, *12*, 487–495, doi:10.1109/JSEN.2011.2107738.

71. Kassal, P.; Steinberg, M.D.; Steinberg, I.M. Wireless chemical sensors and biosensors: A review. *Sens. Actuators B Chem.* **2018**, *266*, 228–245, doi:10.1016/j.snb.2018.03.074.

72. Zhang, D.; Liu, Q. Biosensors and bioelectronics on smartphone for portable biochemical detection. *Biosens. Bioelectron.* **2016**, *75*, 273–284, doi:10.1016/j.bios.2015.08.037.

73. Roda, A.; Michelini, E.; Zangheri, M.; Di Fusco, M.; Calabria, D.; Simoni, P. Smartphone-based biosensors: A critical review and perspectives. *TrAC Trends Anal. Chem.* **2016**, *79*, 317–325, doi:10.1016/j.trac.2015.10.019.

74. Quesada-González, D.; Merkoçi, A. Mobile phone-based biosensing: An emerging "diagnostic and communication" technology. *Biosens. Bioelectron.* **2017**, *92*, 549–562, doi:10.1016/j.bios.2016.10.062.

75. Grossi, M. A sensor-centric survey on the development of smartphone measurement and sensing systems. *J. Meas.* **2019**, *135*, 572–592, doi:10.1016/j.measurement.2018.12.014.

76. Johnston, A.H.; Weiss, G.M. Smartwatch-based biometric gait recognition. In Procedings of the IEEE International Conference on Biometrics Theory, Applications and Systems, Arlington, VA, USA, 8–11 September 2015; doi:10.1109/BTAS.2015.7358794.

77. Xu, W.; Shen, Y.; Zhang, Y.; Bergmann, N.; Hu, W. Gait-Watch: A Context-aware Authentication System for Smart Watch Based on Gait Recognition. In Procedings of the IEEE/ACM International Conference on Internet-of-Things Design and Implementation, Pittsburgh, PA, USA, 18–21 April 2017; ISBN 978-1-4503-4966-6.

78. Bandodkar, A.M.; You, J.M.; Kim, N.H.; Gu, Y.; Kumar, R.; Vinu Mohan, A.M.; Kurniawan, J.; Imani, S.; Nakagawa, T.; Parish, B.; et al. Soft, stretchable, high power density electronic skin-based biofuel cells for scavenging energy from human sweat. *Energy Environ. Sci.* **2017**, *10*, 1581–1589, doi:10.1039/c7ee00865a.

79. Lorwongtragool, P.; Sowade, E.; Watthanawisuth, N.; Baumann, R.R.; Kerdcharoen, T. A Novel Wearable Electronic Nose for Healthcare Based on Flexible Printed Chemical Sensor Array. *Sensors* **2014**, *14*, 19700–19712, doi:10.3390/s141019700.

80. Mostafalu, P.; Lenk, W.; Dokmeci, M.R.; Ziaie, B.; Khademhosseini, A.; Sonkusale, S.R. Wireless Flexible Smart Bandage for Continuous Monitoring of Wound Oxygenation. *IEEE Trans. Biomed. Circuits Syst.* **2015**, *9*, 670–677, doi:10.1109/TBCAS.2015.2488582.

81. Matzeu, G.; O'Quigley, C.; McNamara, E.; Zuliani, C.; Fay, C.; Glennon, T.; Diamond, D. An integrated sensing and wireless communications platform for sensing sodium in sweat. *Anal. Methods* **2016**, *8*, 64–71, doi:10.1039/C5AY02254A.

82. Farooqui, M.F.; Shamim, A. Low Cost Inkjet Printed Smart Bandage for Wireless Monitoring of Chronic Wounds. *Sci. Rep.* **2016**, *6*, 28949, doi:10.1038/srep28949.

83. Kim, J.; Imani, S.; de Araujo, W.R.; Warchall, J.; Valdés-Ramírez, G.; Paixão, T.R.; Mercier, P.P.; Wang, J. Wearable salivary uric acid mouthguard biosensor with integrated wireless electronics. *Biosens. Bioelectron.* **2015**, *74*, 1061–1068, doi:10.1016/j.bios.2015.07.039

84. Kim, J.; Kim, J.; Jeerapan, I.; Imani, S.; Cho, T.N.; Bandodkar, A.; Cinti, S.; Mercier, P.P.; Wang, J. Noninvasive Alcohol Monitoring Using a Wearable Tattoo-Based Iontophoretic-Biosensing System. *ACS Sens.* **2016**, *1*, 11–1019, doi:10.1021/acssensors.6b00356.

85. Yao, S.; Myers, A.; Malhotra, A.; Lin, F.; Bozkurt, A.; Muth, J.F.; Zhu, Y. A Wearable Hydration Sensor with Conformal Nanowire Electrodes. *Adv. Healthc. Mater.* **2017**, *6*, 1601159, doi:10.1002/adhm.201601159.

86. Emaminejad, S.; Gao, W.; Wu, E.; Davies, Z.A.; Nyein, H.Y.Y.; Challa, S.; Ryan, S.P.; Fahad, H.M.; Chen, K.; Shahpar, Z.; et al. Autonomous sweat extraction and analysis applied to cystic fibrosis and glucose monitoring using a fully integrated wearable platform. *PNAS* **2017**, *114*, 4625–4630, doi:10.1073/pnas.1701740114.

87. Oletic, D.; Bilas, V. Design of Sensor Node for Air Quality Crowdsensing. In Proceedings of the IEEE Sensors Applications Symposium, Zadar, Croatia, 13–15 April 2015.

88. Azzarelli, J.M.; Mirica, K.A.; Ravnsbæk, J.B.; Swager, T.M. Wireless gas detection with a smartphone via RF communication. *PNAS* **2014**, *111*, 18162–18166, doi:10.1073/pnas.1415403111.

89. Steinberg, M.D.; Kassal, P.; Kereković, I.; Steinberg, I.M. A wireless potentiostat for mobile chemical sensing and biosensing. *Talanta* **2015**, *143*, 178–183, doi:10.1016/j.talanta.2015.05.028.

90. Kassal, P.; Zubak, M.; Scheipl, G.; Mohr, G.J.; Steinberg, M.D.; Steinberg, I.M. Smart bandage with wireless connectivity for optical monitoring of pH. *Sens. Actuators B Chem.* **2017**, *246*, 455–460, doi:10.1016/j.snb.2017.02.095.

91. Bard, A.J.; Faulkner, L.R. *Electrochemical Methods, Fundamentals and Applications*, 2nd ed.; Wiley: New York, NY, USA, 2011; ISBN 0-471-04372-9.

92. Wang, J. *Analytical Electrochemistry*, 3rd ed.; Wiley: New York, NY, USA, 2006; ISBN 0-471-67879-1.

93. Scholz, F. (Ed.) *Electroanalytical Methods*; Springer: Berlin/Heidelberg, Germany, 2010; ISBN 978-3-642-02914-1.

94. Koryta, J. Ion-Selective Electrodes. *Ann. Rev. Mater. Sci.* **1986**, *16*, 13–27, doi:10.1146/annurev.ms.16.080186.000305.

95. Wolfrum, B.; Zevenbergen, M.; Lemay, S. Nanofluidic Redox Cycling Amplification for the Selective Detection of Catechol. *Anal. Chem.* **2008**, *80*, 972–977, doi:10.1021/ac7016647

96. Goluch, E.D.; Wolfrum, B.; Singh, P.S.; Zevenbergen, M.A.G.; Lemay, S.G. Redox cycling in nanofluidic channels using interdigitated electrodes. *Anal. Bioanal. Chem.* **2010**, *394*, 447–456, doi:10.1007/s00216-008-2575-x.

97. Straver, M.G.; Odijk, M.; Olthuisa, W.; van den Berg, A. A simple method to fabricate electrochemical sensor systems with predictable high-redox cycling amplification. *Lab Chip* **2012**, *12*, 1548–1553, doi:10.1039/C2LC21233A.

98. Barnes, E.O.; Lewis, G.E.M.; Dale, S.E.C.; Marken, F.; Compton, R.G. Generator-collector double electrode systems: A review. *Analyst* **2012**, *137*, 1068, doi:10.1039/C2AN16174E.

99. del Campo, F.J.; Abad Muñoz, Ll.; Illa, X.; Tsai, Y.C. Determination of heterogeneous electron transfer rate constants at interdigitated nanoband electrodes fabricated by an optical mix-and-match process. *Sens. Actuators B Chem.* **2014**, *194*, 86–95, doi:10.1016/j.snb.2013.12.016.

100. Orazem, M.E.; Tribollet, B. *Electrochemical Impedance Spectroscopy*; Wiley: New York, NY, USA, 2008; ISBN 978-0-470-04140-6.

101. Gabrielli, C. *Identification of Electrochemical Processes by Frequency Response Analysis*; Solartron Instrumentation Group: Farnborough, UK; 1980.

102. Tsividis, Y.; Milios, J. A detailed look at electrical equivalents of uniform electrochemical diffusion using nonuniform resistance–capacitance ladders. *J. Electroanal. Chem.* **2013**, *707*, 156–165, doi:10.1016/j.jelechem.2013.08.017.

103. Enz, C.C.; Temes G.C. Circuit techniques for reducing the effects of op-amp imperfections: autozeroing, correlated double sampling, and chopper stabilization. *Proc. IEEE* **1996**, *84*, 1584–1614, doi:10.1109/5.542410.

104. Dei, M.; Bruschi, P.; Piotto, M. Design of CMOS chopper amplifiers for thermal sensor interfacing. In Proceedings of the IEEE PhD Research in Microelectronics and Electronics, Istanbul, Turkey, 22 June–25 April 2008; doi:10.1109/RME.2008.4595761.

105. Dei, M.; Bruschi, P.; Piotto, M. A compact CMOS Gm-C biquadratic cell for chopper amplifier band limiting. In Proceedings of the IEEE PhD Research in Microelectronics and Electronics, Cork, Ireland, 12–17 July 2009; doi:10.1109/RME.2009.5201363.

106. Chandrakumar, H. A 0.6uW/Channel, Frequency Division Multiplexed Amplifier for Neural Recording Systems. Ph.D. Thesis, University of California, Los Angeles, CA, USA, 2012.

107. Schreier, R.; Temes, G.C. *Understanding Delta-Sigma Data Converters*, 1st ed.; Wiley-IEEE Press: Piscataway, NJ, USA, 2005; ISBN 0471465852.

108. Menolfi, C.; Huang, Q. A Fully Integrated, Untrimmed CMOS Instrumentation Amplifier with Submicrovolt Offset. *IEEE J. Solid-State Circuits* **1999**, *34*, 415–450, doi:10.1109/4.748194.

109. Wu, R.; Makinwa, K.A.A.; Huijsing, J.H. A chopper current-feedback instrumentation amplifier with a 1 mHz 1/f noise corner and an AC-coupled ripple reduction loop. *IEEE J. Solid-State Circuits* **2009**, *44*, 3232–3243, doi:10.1109/JSSC.2009.2032710.

110. Fan, Q.; Huijsing, J.H.; Makinwa, K.A.A. A 21 nV/$\sqrt{\text{Hz}}$ chopper-stabilized multi-path current-feedback instrumentation amplifier with 2 μV offset. *IEEE J. Solid-State Circuits* **2012**, *47*, 464–475, doi:10.1109/JSSC.2011.2175269.

111. Kusuda, Y. Auto correction feedback for ripple suppression in a chopper amplifier. *IEEE J. Solid-State Circuits* **2010**, *45*, 1436–1445, doi:10.1109/JSSC.2010.2048142.

112. Chandrakumar, H.; Marcović, D. A Simple Area-Efficient Ripple-Rejection Technique for Chopped Biosignal Amplifiers. *IEEE Trans. Circuits Syst. II* **2015**, *62*, 189–193, doi:10.1109/TCSII.2014.2387686.

113. Bilotti, A.; Monreal, G. Chopper-stabilized amplifiers with a track-and-hold signal demodulator. *IEEE Trans. Circuits Syst. I* **1999**, *46*, 490–495, doi:10.1109/81.754850.

114. Belloni, M.; Bonizzoni, E.; Fornasari, A.; Maloberti, F. A micropower chopper—CDS operational amplifier. *IEEE J. Solid-State Circuits* **2010**, *45*, 2521–2529, doi:10.1109/JSSC.2010.2076430.

115. Pertijs, M.A.P.; Kindt, W.J. A 140 dB-CMRR current-feedback instrumentation amplifier employing ping-pong auto-zeroing and chopping. *IEEE J. Solid-State Circuits* **2010**, *45*, 2044–2056, doi:10.1109/JSSC.2010.2060253.

116. Butti, F.; Bruschi, P.; Dei, M.; Piotto, M. A compact instrumentation amplifier for MEMS thermal sensor interfacing. *Analog Integr. Circuits Signal Process.* **2012**, *72*, 585–594, doi:10.1007/s10470-011-9661-2.

117. Butti, F.; Piotto, M.; Bruschi, P. A chopper instrumentation amplifier with input resistance boosting by means of Synchronous Dynamic Element Matching. *IEEE Trans. Circuits Syst. I Regul. Pap.* **2016**, *64*, 753–764, doi:10.1109/TCSI.2016.2633384.

118. Muller, R.; Gambini, S.; Rabaey, J.M. A 0.013 mm^2, 5 μW, DC-coupled neural signal acquisition IC with 0.5 V supply. *IEEE J. Solid-State Circuits* **2012**, *47*, 232–243, doi:10.1109/JSSC.2011.2163552.

119. Akita, I.; Ishida, M. A chopper-stabilized instrumentation amplifier using area-efficient self-trimming technique. *Analog Integr. Circuits Process.* **2014**, *81*, 571–582, doi:10.1007/s10470-014-0371-4.

120. Sherry, A. *Chopping on* $\Sigma - \Delta$ *ADCs*; Analog Devices–Application Note, AN-609; Norwood, MA, USA, 2003. Available online: https://www.analog.com/media/en/technical-documentation/application-notes/AN-609.pdf (accessed on 30 January 2019).

121. Pertijs, M.A.P.; Makinwa, K.A.A.; Huijsing, J.H. A CMOS Smart Temperature Sensor With a 3σ Inaccuracy of $\pm 0.1\,°C$ From $-55\,°C$ to $125\,°C$. *IEEE J. Solid-State Circuits* **2005**, *40*, 2805–2815, doi:10.1109/JSSC.2005.858476.

122. Catania, A.; Ria, A.; Del Cesta, S.; Piotto, M.; Bruschi, P. Analysis and Simulation of Chopper Stabilization Techniques Applied to Delta-Sigma Converters. In Proceedings of the International Conference on Synthesis, Modeling, Analysis and Simulation Methods and Applications to Circuit Design, Prague, Czech Republic, 2–5 July 2018; doi:10.1109/SMACD.2018.8434904.

123. Zhao, Y.S.; Tang, S.K.; Ko, C.T.; Pun, K.P. A chopper-stabilized high-pass Delta–Sigma Modulator with reduced chopper charge injection. *Microelectron. J.* **2011**, *42*, 733–739, doi:10.1016/j.mejo.2011.02.002.

124. Wang, H.; Wang, X.; Barfidokht, A.; Park, J.; Wang, J.; Mercier, P.P. A Battery-Powered Wireless Ion Sensing System Consuming 5.5 nW of Average Power. *IEEE J. Solid-State Circuits* **2018**, *53*, 2043–2053, doi:10.1109/JSSC.2018.2815657.

125. Craninckx, J.; van der Plas, G. A 65fJ/Conversion-Step 0-to-50MS/s 0-to-0.7mW 9b Charge-Sharing SAR ADC in 90nm Digital CMOS. In Proceedings of the IEEE International Solid-State Circuits Conference, San Francisco, CA, USA, 11–15 February 2007; doi:10.1109/ISSCC.2007.373386.

126. Wang H.; Mercier P.P. A Reference-Free Capacitive-Discharging Oscillator Architecture Consuming 44.4 pW/75.6 nW at 2.8 Hz/6.4 kHz. *IEEE J. Solid-State Circuits* **2011**, *51*, 1423–1435, doi:10.1109/JSSC.2016.2554883.

127. Harpe, P.J.A.; Zhou, C.; Bi, Y.; van der Meijs, N.P.; Wang, X.; Philips, K.; Dolmans, G.; de Groot, H. A 26 μW 8 bit 10 MS/s Asynchronous SAR ADC for Low Energy Radios. *IEEE J. Solid-State Circuits* **2011**, *46*, 1585–1595, doi:10.1109/JSSC.2011.2143870.

128. Ghoreishizadeh, S.S.; Baj-Rossi, C.; Cavallini, A.; Carrara, S.; De Micheli, G. An Integrated Control and Readout Circuit for Implantable Multi-Target Electrochemical Biosensing. *IEEE Trans. Biomed. Circuits Syst.* **2014**, *8*, 891–898, doi:10.1109/TBCAS.2014.2315157.

129. Xiao, Z.; Tan, X.; Chen, X.; Chen, S.; Zhang, Z.; Zhang, H.; Wang, J.; Huang, Y.; Zhang, P.; Zheng, L.; Min, H. An Implantable RFID Sensor Tag toward Continuous Glucose Monitoring. *IEEE J. Biomed. Health Inform.* **2015**, *19*, 910–919, doi:10.1109/JBHI.2015.2415836.

130. Zuo, L.; Islam, S.K.; Mahbub, I.; Quaiyum, F. A Low-Power 1-V Potentiostat for Glucose Sensors. *IEEE Trans. Circuits Syst. II Exp. Briefs* **2015**, *62*, 204–208, doi:10.1109/TCSII.2014.2387691.

131. Ghodsevali, E.; Morneau-Gamache, S.; Mathault, J.; Landari, H.; Boisselier, É.; Boukadoum, M.; Gosselin, B.; Miled, A. Miniaturized FDDA and CMOS Based Potentiostat for Bio-Applications. *Sensors* **2017**, *17*, 810, doi:10.3390/s17040810.

132. Martin, S.M.; Gebara, F.H.; Strong, T.D.; Brown, R.B. A Fully Differential Potentiostat. *IEEE Sens. J.* **2009**, *6*, 135–142, doi:10.1109/JSEN.2008.2011085.

133. Giagkoulovits, C.; Cheah, B.C.; Al-Rawhani, M.A.; Accarino, C.; Busche, C.; Grant, J.P.; Cumming, D.R.S. A 16 × 16 CMOS amperometric microelectrode array for simultaneous electrochemical measurements. *IEEE Trans. Circuits Syst. I Regul. Pap.* **2018**, *65*, 2821–2831, doi:10.1109/TCSI.2018.2794502.

134. Li, H.; Parsnejad, S.; Ashoori, E.; Thompson, C.; Purcell, E.K.; Mason, A.J. Ultracompact Microwatt CMOS Current Readout With Picoampere Noise and Kilohertz Bandwidth for Biosensor Arrays. *IEEE Trans. Biomed. Circuits Syst.* **2017**, *12*, 35–46, doi:10.1109/TBCAS.2017.2752742.

135. Crescentini, M.; Bennati, M.; Carminati, M.; Tartagni, M. Noise Limits of CMOS Current Interfaces for Biosensors: A Review. *IEEE Trans. Biomed. Circuits Syst.* **2014**, *8*, 278–292, doi:10.1109/TBCAS.2013.2262998.

136. Ahmadi, M.M.; Jullien, G.A. Current-Mirror-Based Potentiostats for Three-Electrode Amperometric Electrochemical Sensors. *IEEE Trans. Circuits Syst. I Regul. Pap.* **2009**, *56*, 1339–1348, doi:10.1109/TCSI.2008.2005927.

137. Al Mamun, K.A.; Islam, S.K.; Hensley, D.K.; McFarlane, N. A Glucose Biosensor Using CMOS Potentiostat and Vertically Aligned Carbon Nanofibers. *IEEE Trans. Biomed. Circuits Syst.* **2016**, *10*, 807–816, doi:10.1109/TBCAS.2016.2557787.

138. Lin, F.; Lu, S.; Liao, Y. A 2.2μW, -12 dBm RF-Powered Wireless Current Sensing Readout Interface IC With Injection-Locking Clock Generation. *IEEE Trans. Circuits Syst. I Regul. Pap.* **2016**, *63*, 950–959, doi:10.1109/TCSI.2016.2546398.

139. Tsai, J.; Kuo, C.; Lin, S.; Lin, F.; Liao, Y. A Wirelessly Powered CMOS Electrochemical Sensing Interface With Power-Aware RF-DC Power Management. *IEEE Trans. Circuits Syst. I Regul. Pap.* **2018**, *65*, 2810–2820, doi:10.1109/TCSI.2018.2797238.

140. Razavi, B. *Design of Analog CMOS Integrated Circuits*; McGraw-Hill: New York, NY, USA, 2001; ISBN 0072380322.

141. Levine, P.M.; Gong, P.; Levicky, R.; Shepard, K.L. Active CMOS Sensor Array for Electrochemical Biomolecular Detection. *IEEE J. Solid-State Circuits* **2008**, *43*, 1859–1871, doi:10.1109/JSSC.2008.925407.

142. Yang, C.; Huang, Y.; Hassler, B.L.; Worden, R.M.; Mason, A.J. Amperometric Electrochemical Microsystem for a Miniaturized Protein Biosensor Array. *IEEE Trans. Biomed. Circuits Syst.* **2009**, *3*, 160–168, doi:10.1109/TBCAS.2009.2015650.

143. Nazari, M.H.; Mazhab-Jafari, H.; Leng, L.; Guenther, A.; Genov, R. CMOS Neurotransmitter Microarray: 96-Channel Integrated Potentiostat With On-Die Microsensors. *IEEE Trans. Biomed. Circuits Syst.* **2013**, *7*, 338–348, doi:10.1109/TBCAS.2012.2203597.

144. Jafari, H.M.; Abdelhalim, K.; Soleymani, L.; Sargent, E.H.; Kelley, S.O.; Genov, R. Nanostructured CMOS Wireless Ultra-Wideband Label-Free PCR-Free DNA Analysis SoC. *IEEE J. Solid-State Circuits* **2014**, *49*, 1223–1241, doi:10.1109/JSSC.2014.2312571.

145. Liu, Y.T.; Chen, M.; Li, Z.C.; Wang, Y.; Chen, J. A High Dynamic Range Analog-front-end IC for Electrochemical Amperometric and Voltammetric Sensor. *Microelectron. J.* **2015**, *46*, 716–722, doi:10.1016/j.mejo.2015.05.009.

146. Ghoreishizadeh, S.S.; Taurino, I.; De Micheli, G.; Carrara, S.; Georgiou, P. A Differential Electrochemical Readout ASIC With Heterogeneous Integration of Bio-Nano Sensors for Amperometric Sensing. *IEEE Trans. Biomed. Circuits Syst.* **2016**, *11*, 1148–1159, doi:10.1109/TBCAS.2017.2733624.

147. Yin, H.; Mu, X.; Li, H.; Liu, X.; Mason, A.J. CMOS Monolithic Electrochemical Gas Sensor Microsystem Using Room Temperature Ionic Liquid. *IEEE Sens. J.* **2018**, *18*, 7899–7906, doi:10.1109/JSEN.2018.2863644.

148. Murmann, B. Thermal Noise in Track-and-Hold Circuits: Analysis and Simulation Techniques. *IEEE Solid State Circuits Mag.* **2012**, *4*, 46–54, doi:10.1109/MSSC.2012.2192190.

149. Hasting, A. *The Art of Analog Layout*; Pearson Prentice Hall: Upper Saddle River, NJ, USA, 2006; ISBN 0131464108.

150. Tan, X.; Chen, X.; Zhang, Y.; Wang, J.; Min, H.; Huang, Y.; Mason, A.J. 4.3 µW 10 fA-sensitivity dual-mode current converter for implantable glucose monitoring. *Electron. Lett.* **2015**, *51*, 1484–1486, doi:10.1049/el.2015.2238.

151. Dei, M.; Sacristán, J.; Marigó, E.; Soundara, M.; Terés, L.; Serra-Graells, F. A 10-bit Linearity Current-Controlled Ring Oscillator with Rolling Regulation for Smart Sensing. In Procedings of the IEEE International Symposium on Circuits and Systems, Baltimore, MD, USA, 28–31 May 2017; doi:10.1109/ISCAS.2017.8050228.

152. Massicotte, G.; Carrara, S.; Di Micheli, G.; Sawan, M. A CMOS Amperometric System for Multi-Neurotransmitter Detection. *IEEE Trans. Biomed. Circuits Syst.* **2016**, *10*, 731–741, doi:10.1109/TBCAS.2015.2490225.

153. Dai, S.; Perera, R.T.; Yang, Z.; Rosenstein, J.K. A 155-dB Dynamic Range Current Measurement Front End for Electrochemical Biosensing. *IEEE Trans. Biomed. Circuits Syst.* **2016**, *10*, 935–944, doi:10.1109/TBCAS.2016.2612581.

154. Dei, M.; Figueras, R.; Margarit, J.M.; Terés, L.; Serra-Graells, F. Highly Linear Integrate-and-Fire Modulators with Soft Reset for Low-Power High-Speed Imagers. In Procedings of the IEEE International Symposium on Circuits and Systems, Baltimore, MD, USA, 28–31 May 2017; doi:10.1109/ISCAS.2017.8050411.

155. Stanacevic, M.; Murari, K.; Rege, A.; Cauwenberghs, G.; Thakor, N.V. VLSI Potentiostat Array With Oversampling Gain Modulation for Wide-Range Neurotransmitter Sensing. *IEEE Trans. Biomed. Circuits Syst.* **2007**, *1*, 63–72, doi:10.1109/TBCAS.2007.893176.

156. Sutula, S.; Pallarés Cuxart, J.; Gonzalo-Ruiz, J.; Muñoz-Pascual, F.X.; Terés, L.; Serra-Graells, F. A 25-µW All-MOS Potentionstatic Delta-Sigma ADC for Smart Electrochemical Sensors. *IEEE Trans. Circuits Syst. I Regul. Pap.* **2014**, *61*, 671–679, doi:10.1109/TCSI.2013.2284179.

157. Li, H.; Boling, C.S.; Mason, A.J. CMOS Amperometric ADC With High Sensitivity, Dynamic Range and Power Efficiency for Air Quality Monitoring. *IEEE Trans. Biomed. Circuits Syst.* **2016**, *10*, 817–827, doi:10.1109/TBCAS.2016.2571306.

158. Aymerich, J.; Dei, M.; Terés, L.; Serra-Graells, F. Design of a Low-Power Potentiostatic Second-Order CT Delta-Sigma ADC for Electrochemical Sensors. In Proceedings of the IEEE PhD Research in Microelectronics and Electronics, Taormina, Italy, 12–15 June 2017; doi:10.1109/PRIME.2017.7974118.

159. Aymerich, J.; Dei, M.; Terés, L.; Serra-Graells, F. A 6.5-µW 70-dB 0.18-µm CMOS Potentiostatic Delta-Sigma for Electrochemical Sensors. In Proceedings of the IEEE PhD Research in Microelectronics and Electronics, Prague, Czech Republic, 2–5 July 2018; doi:10.1109/PRIME.2018.8430329.

160. Aymerich, J.; Márquez, A.; Terés, L.; Muñoz-Berbel, X.; Jiménez, C.; Domínguez, C.; Serra-Graells, F.; Dei, M. Cost-effective smartphone-based reconfigurable electrochemical instrument for alcohol determination in whole blood samples. *Biosens. Bioelectron.* **2018**, *117*, 736–742, doi:10.1016/j.bios.2018.06.044.

161. Ghanbari, S.; Habibi, M.; Magierowski, S. A High-Efficiency Discrete Current Mode Output Stage Potentiostat Instrumentation for Self-Powered Electrochemical Devices. *IEEE Trans. Instrum. Meas.* **2018**, *67*, 2247–2255, doi:10.1109/TIM.2018.2811450.

162. Razavi, B. The StrongARM Latch. *IEEE Solid State Circuits Mag.* **2015**, 12–17, doi:10.1109/MSSC.2015.2418155.

163. Bergveld, P. Development, Operation, and Application of the Ion-Sensitive Field-Effect Transistor as a Tool for Electrophysiology. *IEEE Trans. Biomed. Eng.* **1972**, *5*, 342–352, doi:10.1109/TBME.1972.324137.

164. Bergveld, P. Thirty years of ISFETOLOGY: What happened in the past 30 years and what may happen in the next 30 years. *Sens. Actuator B Chem.* **2003**, *88*, 1–20, doi:10.1016/S0925-4005(02)00301-5.

165. Bausells, J.; Carrabina, J.; Errachid, A.; Merlos, A. Ion-sensitive field-effect transistors fabricated in a commercial CMOS technology. *Sens. Actuator B Chem.* **2003**, *57*, 56–62, doi:10.1016/S0925-4005(99)00135-5.

166. The International Technology Roadmap for Semiconductors (ITRS) 2.0, "More Moore". 2015. Available online: http://www.itrs2.net (accessed on 21 January 2019).

167. Nakata, S.; Arie, T.; Akita, S.; Takei, K. Wearable, Flexible, and Multifunctional Healthcare Device with an ISFET Chemical Sensor for Simultaneous Sweat pH and Skin Temperature Monitoring. *ACS Sens.* **2017**, *2*, 443–448, doi:10.1021/acssensors.7b00047.

168. van Hal, R.E.G.; Eijkel, J.C.T.; Bergveld, P. A novel description of ISFET sensitivity with the buffer capacity and double-layer capacitance as key parameters. *Sens. Actuator B Chem.* **1995**, *24*, 201–205, doi:10.1016/0925-4005(95)85043-0.

169. Georgiou, P.; Toumazou, C. ISFET characteristics in CMOS and their application to weak inversion operation. *Sens. Actuator B Chem.* **2009**, *143*, 211–217, doi:10.1016/j.snb.2009.09.018.

170. Moser, N.; Lande, T.S.; Toumazou, C.; Georgiou, P. ISFETs in CMOS and Emergent Trends in Instrumentation: A Review. *IEEE Sens. J.* **2016**, *16*, 6496–6514, doi:10.1109/JSEN.2016.2585920.

171. Hanazato, Y.; Nakako, M.; Shiono, S.; Maeda, M. Integrated multi-biosensors based on an ion-sensitive field-effect transistor using photolithographic techniques. *IEEE Trans. Electron Devices* **1989**, *36*, 1303–1310, doi:10.1109/16.30936.

172. Enz, C.C.; Krummenacher, F.; Vittoz E.A. An Analytical MOS Transistor Model Valid in All Regions of Operation and Dedicated to Low-Voltage and Low-Current Applications. *Analog Integr. Circuits Signal Process.* **1995**, *8*, 83–114, doi:10.1007/BF01239381.

173. Jimenez, C.; Orozco, J.; Baldi, A. ISFET based sensors: Fundamentals and applications. *Encycl. Sens.* **2006**, *5*, 151–196.

174. Miscourides, N.; Yu, L.; Rodriguez-Manzano, J.; Georgiou, P. A 12.8 k Current-Mode Velocity-Saturation ISFET Array for On-Chip Real-Time DNA Detection. *IEEE Trans. Biomed. Circuits Syst.* **2018**, *12*, 1202–1214, doi:10.1109/TBCAS.2018.2851448.

175. Moser, N.; Rodriguez-Manzano, J.; Lande, T.S.; Georgiou, P. A Scalable ISFET Sensing and Memory Array With Sensor Auto-Calibration for On-Chip Real-Time DNA Detection. *IEEE Trans. Biomed. Circuits Syst.* **2018**, *12*, 390–401, doi:10.1109/TBCAS.2017.2789161.

176. Alhoshany, A.; Sivashankar, S.; Mashraei, Y.; Omran, H.; Salama, K.N. A Biosensor-CMOS Platform and Integrated Readout Circuit in 0.18-μm CMOS Technology for Cancer Biomarker Detection. *Sensors* **2017**, *17*, 1942, doi:10.3390/s17091942

177. Laborde, C.; Pittino, F.; Verhoeven, H.A.; Lemay, S.G.; Selmi, L.; Jongsma, M.A.; Widdershoven, F.P. Real-time imaging of microparticles and living cells with CMOS nanocapacitor arrays. *Nat. Nanotechnol.* **2015**, *10*, 791, doi:10.1038/NNANO.2015.163.

178. Segura-Quintano, F.; Sacristán-Riquelme, J.; García-Cantón, J.; Osés M.T.; Baldi, A. Towards Fully Integrated Wireless Impedimetric Sensors. *Sensors* **2010**, *10*, 4071–4082, doi:10.3390/s100404071.

179. Rodriguez, S.; Ollmar, S.; Waqar, M.; Rusu, A. A Batteryless Sensor ASIC for Implantable Bio-Impedance Applications. *IEEE Trans. Biomed. Circuits Syst.* **2015**, *10*, 533–544, doi:10.1109/TBCAS.2015.2456242.

180. Xu, J.; Harpe, P.; Pettine, J.; Van Hoof, C.; Yazicioglu, R.F. A low power configurable bio-impedance spectroscopy (BIS) ASIC with simultaneous ECG and respiration recording functionality. In Proceedings of the European Solid-State Circuits Conference, Graz, Austria, 14–18 September 2015; doi:10.1109/ESSCIRC.2015.7313911.

181. Jafari, H.; Soleymani, L.; Genov, R. 16-Channel CMOS Impedance Spectroscopy DNA Analyzer With Dual-Slope Multiplying ADCs. *IEEE Trans. Biomed. Circuits Syst.* **2012**, *6*, 468–478, doi:10.1109/TBCAS.2012.2226334.

182. Manickam, A.; Chevalier, A.; McDermott, M.; Ellington, A.D.; Hassibi, A. A CMOS Electrochemical Impedance Spectroscopy (EIS) Biosensor Array. *IEEE Trans. Biomed. Circuits Syst.* **2010**, *4*, 379–390, doi:10.1109/TBCAS.2010.2081669.

183. Min, M.; Parve, T. Improvement of lock-in electrical bio-impedance analyzer for implantable medical devices. *IEEE Trans. Instrum. Meas.* **2007**, *56*, 968–974, doi:10.1109/TIM.2007.894172.

184. Wei, C.; Wang, Y.; Liu, B. Wide-range filter-based sinusoidal wave synthesizer for electrochemical impedance spectroscopy measurements. *IEEE Trans. Biomed. Circuits Syst.* **2014**, *8*, 442–450, doi:10.1109/TBCAS.2013.2274100.

185. Crescentini, M.; Bennati, M.; Tartagni, M. Recent trends for (bio) chemical impedance sensor electronic interfaces. *Electroanal* **2012**, *24*, 563–572, doi:10.1002/elan.201100547.

186. Qu, G.; Wang, H.; Zhao, Y.; O'Donnell, J.; Lyden, C.; Liu, Y.; Ding, J.; Dempsey, D.; Chen, L.; Bourke, D.; Gu, S.; Gao, J.; Lu, L.; Wang, L.; Li, X.; Li, H.; Chu, C.; Yang, L. A 0.28 mΩ-sensitivity 105 dB-dynamic-range electrochemical impedance spectroscopy soc for electrochemical gas detection. In Proceedings of the International Solid-State Circuits Conference, San Francisco, CA, USA, 11–15 February 2018; doi:10.1109/ISSCC.2018.8310296.

187. Chen, T.; Wu, W.; Wei, C.; Darling, R.B.; Liu, B. Novel 10-Bit Impedance-to-Digital Converter for Electrochemical Impedance Spectroscopy Measurements. *IEEE Trans. Biomed. Circuits Syst.* **2016**, *11*, 370–379, doi:10.1109/TBCAS.2016.2592511.

188. Shleev, S.; Bergel, A.; Gorton, L. Biological fuel cells: Divergence of opinion. *Bioelectrochemistry*, **2015**, *106*, 1–2, doi:10.1016/j.bioelechem.2015.07.006.

189. Katz, E.; Bückmann, A.F.; Willner, I. Self-Powered Enzyme-Based Biosensors. *J. Am. Chem. Soc.* **2001**, *123*, 10752–10753, doi:10.1021/ja0167102.

190. Jeerapan, I.; Sempionatto, J.R.; Pavinatto, A.; Youa, J.M.; Wang, J. Stretchable biofuel cells as wearable textile-based self-powered sensors. *J. Mater. Chem. A* **2016**, *4*, 18342–18353, doi:10.1039/c6ta08358g.

191. Yeknami, A.F.; Wang, X.; Jeerapan, I.; Imani, S.; Nikoofard, A.; Wang, J.; Mercier, P.P. A 0.3-V CMOS Biofuel-Cell-Powered Wireless Glucose/Lactate Biosensing System. *IEEE J. Solid-State Circuits* **2018**, doi:10.1109/JSSC.2018.2869569.

192. Niitsu, K.; Kobayashi, A.; Nishio, Y.; Hayashi, K.; Ikeda, K.; Ando, T.; Ogawa, Y.; Kai, H.; Nishizawa, M.; Nakazato, K. A Self-Powered Supply-Sensing Biosensor Platform Using Bio Fuel Cell and Low-Voltage, Low-Cost CMOS Supply-Controlled Ring Oscillator With Inductive-Coupling Transmitter for Healthcare IoT. *IEEE Trans. Circuits Syst. I Regul. Pap.* **2018**, *65*, 2784–2796, doi:10.1109/TCSI.2018.2791516.

MDPI
St. Alban-Anlage 66
4052 Basel
Switzerland
Tel. +41 61 683 77 34
Fax +41 61 302 89 18
www.mdpi.com

Electronics Editorial Office
E-mail: electronics@mdpi.com
www.mdpi.com/journal/electronics